Design of Coastal Structures and Sea Defenses

Series on Coastal and Ocean Engineering Practice

Series Editor: Young C Kim
(California State University, USA)

Vol. 1: Coastal and Ocean Engineering Practice
edited by Young C Kim

Vol. 2: Design of Coastal Structures and Sea Defenses
edited by Young C Kim

Series on Coastal and Ocean Engineering Practice – Vol. 2

Design of Coastal Structures and Sea Defenses

Editor

Young C Kim

California State University, Los Angeles, USA

NEW JERSEY • LONDON • SINGAPORE • BEIJING • SHANGHAI • HONG KONG • TAIPEI • CHENNAI

Published by

World Scientific Publishing Co. Pte. Ltd.
5 Toh Tuck Link, Singapore 596224
USA office: 27 Warren Street, Suite 401-402, Hackensack, NJ 07601
UK office: 57 Shelton Street, Covent Garden, London WC2H 9HE

Library of Congress Cataloging-in-Publication Data
Kim, Young C., 1936–
Design of coastal structures and sea defenses / Young C Kim (California State University, Los Angeles, USA).
pages cm. -- (Series on coastal and ocean engineering practice ; vol. 2)
Includes bibliographical references and index.
ISBN 978-9814611008 (hardcover : alk. paper)
1. Shore protection. 2. Coastal engineering. 3. Coastal zone management. I. Title.
TC330.K56 2014
627'.58--dc23

2014026826

British Library Cataloguing-in-Publication Data
A catalogue record for this book is available from the British Library.

Cover Picture: Construction of the Main Breakwater of the New Outer Port at Punto Langosteira, La Coruña, Spain.
Reproduced with permission from the Port Authority of La Coruña, Spain.

Printed in Singapore

Preface

Coastal structures are an important component in any coastal protection scheme. They directly control wave and storm surge action or stabilize a beach which provides protection on the coast. This book, Design of Coastal Structures and Sea Defenses, Series on Coastal and Ocean Engineering Practice, Vol. 2, provides the most up-to-date technical advances on the design and construction of coastal structures and sea defenses.

Written by renowned practicing coastal engineers, this edited volume focuses on the latest technology applied in planning, design and construction, effective engineering methodology, unique projects and problems, design and construction challenges, and other lessons learned.

In addition, unique practice in planning, design, construction, maintenance, and performance of coastal and ocean projects will be explored.

Many books have been written about the theoretical treatment of coastal and ocean structures and sea defenses. Much less has been written about the practical aspect of ocean structures and sea defenses. This comprehensive book fills the gap. It is an essential source of reference for professionals and researchers in the areas of coastal, ocean, civil, and geotechnical engineering.

I would like to express my indebtedness to 19 authors and co-authors who have contributed to this series. It has been more than a year long project, and their support and sacrifices have been deeply appreciated. I am very grateful.

Finally, I wish to express my deep appreciation to Mr. Steven Patt of World Scientific Publishing who gave me invaluable support and encouragement from the inception of this series to its realization.

Young C. Kim
Los Angeles, California
May 2014

The Editor

Young C. Kim, Ph.D., Dist.D.CE., F.ASCE, is currently a Professor of Civil Engineering, Emeritus at California State University, Los Angeles. Other academic positions held include a Visiting Scholar of Coastal Engineering at the University of California, Berkeley (1971); a NATO Senior Fellow in Science at the Delft University of Technology in the Netherlands (1975); and a Visiting Scientist at the Osaka City University for the National Science Foundations' U.S.-Japan Cooperative Science Program (1976); and a Visiting Professor, Polytech Nice-Sophia, Universite de Nice-Sophia Antipolis (2011, 2012, and 2013). For more than a decade, he served as Chair of the Department of Civil Engineering (1993-2005) and was Associate Dean of Engineering in 1978. For his dedicated teaching and outstanding professional activities, he was awarded the university-wide Outstanding Professor Award in 1994.

Dr. Kim was a consultant to the U.S. Naval Civil Engineering Laboratory in Port Hueneme and became a resident consultant to the Science Engineering Associates where he investigated wave forces on the Howard-Doris platform structure, now being placed in Ninian Field, North Sea.

Dr. Kim is the past Chair of the Executive Committee of the Waterway, Port, Coastal and Ocean Division of the American Society of Civil Engineering (ASCE). Recently, he served as Chair of the Nominating Committee of the International Association for Hydo-Environment Engineering and Research (IAHR). Since 1998, he served on the International Board of Directors of the Pacific Congress on Marine Science and Technology (PACON). He is the past President of PACON.

Dr. Kim has been involved in organizing 14 national and international conferences, has authored six books, and published 55 technical papers in various engineering journals. Recently, he served as an editor for the Handbook of Coastal and Ocean Engineering which was published by the World Scientific Publishing Company in 2010. In 2011, he was inducted as Distinguished Diplomate of Coastal Engineering from the Academy of Coastal, Ocean, Port and Navigation Engineers (ACOPNE). In 2012, he was elected Fellow of the American Society of Civil Engineers.

Contributors

Hans F. Burcharth
Port and Coastal Engineering Consultant
Professor
Aalborg University, Denmark
hansburcharth@gmail.com

Jang-Won Chae
Emeritus Senior Research Fellow
Coastal Development and Ocean Energy Research Division
Korea Institute of Ocean Science and Technology
Ansan, Korea
jwchae@kiost.ac

Byung Ho Choi
Professor
Department of Civil and Environmental Engineering
Sungkyunkwan University
Suwon, Korea
bhchoi.skku@gmail.com

Jae Cheon Choi
Manager, Civil Engineering Team
Daewoo Engineering and Construction Company, Ltd
Seoul, Korea
Jaecheon.choi@daewooenc.com

Minoru Hanzawa
Director, Technical Research Institute
Fudo Tetra Corporation
Tsuchiura, Ibaraki, Japan
minoru.hanzawa@fudotetra.co.jp

Haiqing Liu Kaczkowski
Senior Coastal Engineer
Coastal Science & Engineering Inc.
Columbia, South Carolina, USA
hkaczkowski@coastalscience.com

Timothy W. Kana
Principal Coastal Scientist
Coastal Science & Engineering Inc.
Columbia, South Carolina, USA
tkana@coastalscience.com

Andrew B. Kennedy
Associate Professor
Department of Civil and Environmental Engineering and Earth Sciences
University of Notre Dame
Notre Dame, Indiana, USA
Andrew.kennedy@nd.edu

Kyeong Ok Kim
Senior Research Scientist
Marine Environments and Conservation Research Division
Korea Institute of Ocean Science and Technology
Ansan, Korea
kokim@kiost.ac

Miguel A. Losada
Professor
IISTA, Universidad de Granada
Granada, Spain
mlosada@ugr.es

Enrique Maciñeira
Port Planning and Strategy Manager
Port Authority of La Coruña
La Coruña, Spain
emacine@puertocoruna.com

Fernando Noya
Port Infrastructure Manager
Port Authority of La Coruña
La Coruña, Spain
fnoya@puertocoruna.com

Woo-Sun Park
Principal Research Scientist
Coastal Development and Ocean Energy Research Division
Korea Institute of Ocean Science and Technology
Ansan, Korea
wspark@kiost.ac

Ken-ichiro Shimosako
Director, Coastal and Ocean Engineering Research Field
Port and Airport Research Institute
Yokosuka, Japan
Shimosako@pari.go.jp

Jane McKee Smith
Research Hydraulic Engineer
Coastal and Hydraulics Laboratory
U.S. Army Engineer Research and Development Center
Vicksburg, Mississippi, USA
Jane.M.Smith@usace.army.mil

Sebastián Solari
Ph.D. Research Assistant
IMFIA Universidad de la Republica
Montevideo, Uruguay
ssolari@fing.edu.uy
ssolari@ugr.es

Alexandros A. Taflanidis
Associate Professor
Department of Civil and Environmental Engineering and Earth Sciences
University of Notre Dame
Notre Dame, Indiana, USA
a.taflanidis@nd.edu

Shigeo Takahashi
President
Port and Airport Research Institute
Yokosuka, Japan
takahashi_s@pari.go.jp

Steven B. Traynum
Coastal Scientist
Coastal Science & Engineering Inc.
Columbia, South Carolina, USA
straynum@coastalscience.com

Jentsje W. van der Meer
Principal
Van der Meer Consulting BV
Akkrum, The Netherlands
Professor of Coastal Structures and Ports
UNESCO-IHE
Delft, The Netherlands
jm@vandermeerconsulting.nl

Contents

CHAPTER 1

SIMULATORS AS HYDRAULIC TEST FACILITIES AT DIKES AND OTHER COASTAL STRUCTURES

Jentsje W. van der Meer

Principal, Van der Meer Consulting BV,
Professor UNESCO-IHE, Delft, The Netherlands
P.O. Box 11, 8490 AA, Akkrum, The Netherlands
E-mail: jm@vandermeerconsulting.nl

The first part of this chapter gives a short description of wave processes on a dike, on what we know, including recent new knowledge. These wave processes are wave impacts, wave run-up and wave overtopping. The second part focuses on description of three Simulators, each of them simulating one of the wave processes and which have been and are being used to test the strength of grass covers on a dike under severe storm conditions. Sometimes they are also applied to measure wave impacts by overtopping wave volumes.

1. Introduction

When incident waves reach a coastal structure such as dike or levee, they will break if the slope is fairly gentle. This may cause impacts on the slope in zone 2, see Figure 1. When large waves attack such a dike the seaward side in this area will often be protected by a placed block revetment or asphalt. The reason is simple: grass covers cannot withstand large wave impacts, unless the slope is very mild.

Above the impact zone the wave runs up the slope and then rushes down the slope till it meets the next up-rushing wave. This is the run-up and run-down zone on the seaward slope (zone 3 in Figure 1). Up-rushing waves that reach the crest will overtop the structure and the flow is only to one side: down the landward slope, see zone's 4 and 5 in Figure 1.

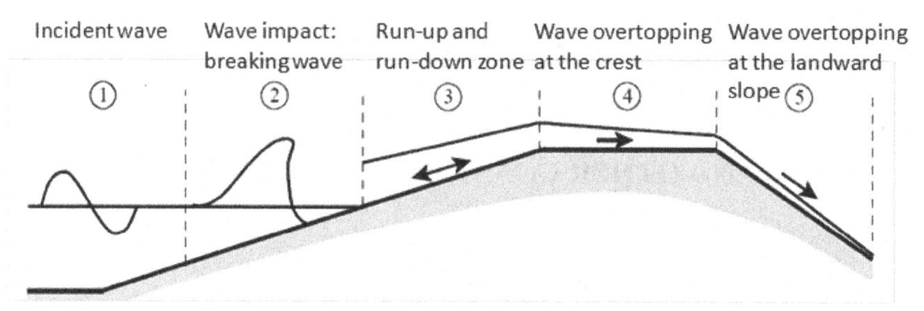

Figure 1. Process of wave breaking, run-up and overtopping at a dike (figure partly from Schüttrumpf (2001)).

Design of coastal structures is often focussed on design values for certain parameters, like the $p_{max,2\%}$ or p_{max} for a design impact pressure, $Ru_{2\%}$ for a wave run-up level and q as mean overtopping discharge or V_{max} as maximum overtopping wave volume. A structure can then be designed using the proper partial safety factors, or with a full probabilistic approach. For wave flumes and wave basins, the waves and the wave processes during wave-structure interaction are simulated correctly using a Froude scale and it are these facilities that have provided the design formulae for the parameters described above.

Whether the strength of coastal structures can also be modelled on small scale depends on the structure considered. The erosion of grass on clay cannot be modelled on a smaller scale and one can only perform resistance testing on real dikes, or on parts moved to a large-scale facility as the Delta Flume of Deltares, The Netherlands, or the GWK in Hannover, Germany. Resistance testing on real dikes can also be performed by the use of Simulators, which is the subject of this chapter. Each Simulator has been developed to simulate only one of the processes in Figure 1 and for this reason three different types of simulator are available today.

If one wants to simulate one of these processes at a real dike, without a wave flume or wave basin, one first has to describe and model the process that should be simulated. Description of the wave-structure-interaction process is, however, much more difficult than just the determination of a design value. The whole process for each wave should be described as good as possible.

2. Simulation of Wave Structure Interaction Processes

2.1. *General aspects*

Three different wave-structure-interaction processes are being recognized on a sloping dike, each with design parameters, but also with other parameters that have to be described for all individual waves. An overall view is given below.

Impacts: *Design parameters:* $p_{max,\,2\%}$; p_{max}
Description of process: distribution of impact pressures, rise times, impact durations, impact width ($B_{impact,50\%}$) and impact locations;

Wave run-up and run-down: *Design parameters:* $Ru_{2\%}$; $Rd_{2\%}$
Description of process: distributions of run-up and run-down levels, velocities along the slope for each wave;

Wave overtopping: *Design parameters:* q; V_{max}
Description of process: distributions of individual overtopping wave volumes, flow velocities, thicknesses and overtopping durations.

2.2. *Wave impacts*

A lot of information on wave impacts has been gathered for the design of placed block revetments on sloping dikes. Klein Breteler [2012] gives a full description of wave impacts and a short summary of the most important parameters is given here. Wave impacts depend largely on the significant wave height. For grassed slopes on a dike the wave impact is often limited, say smaller than H_s = 1 m, otherwise the slope would not be able to resist the impacts. Tests from the Delta Flume with a wave height of about 0.75 m have been used to describe the process of wave impacts. The 2%-value of the maximum pressure can be described by [Klein Breteler, 2012]:

$$\left(\frac{p_{max,2\%}}{\gamma_{berm,p_{max}} H_s}\right)\left(\frac{\rho_w g H_s^2}{\sigma_w}\right)^{0.1} = 12 - 0.28\frac{\xi_{op}}{tan\alpha_T}$$

for $3 \leq \frac{\xi_{op}}{tan\alpha_T} \leq 24$ (1)

with $\gamma_{berm,p_{max}} = 0.17\left(\frac{h_b}{H_s} - 1.2)\right)^2 + 1$

where:

g	=	acceleration of gravity [m/s^2]
H_s	=	significant wave height [m]
h_b	=	vertical distance from swl to berm (positive if berm above swl) [m]
$p_{max,\ x\%}$	=	value which is exceeded by x% of the number of wave impacts related to the number of waves [m water column]
α_T	=	slope angle [°]
$\gamma_{berm,\ pmax}$	=	influence factor for the berm [-]
ρ_w	=	density of water [kg/m^3]
σ_w	=	surface tension [0.073 N/m^2]
ξ_{op}	=	breaker parameter using the peak period T_p [-]

The tests in the Delta Flume clearly showed that the distribution of p is Rayleigh distributed, see Figure 2. The graph has the horizontal axis according to a Rayleigh distribution and a more or less straight line then indicates a Rayleigh distribution. This is indeed the case in Figure 2.

Each parameter can be given as a distribution or exceedance curve, but often the relationship between two parameters is not so straight forward. Figure 3 shows the relationship between the peak pressure and the corresponding width of the impact, $B_{impact,\ 50\%}$ for a wave field with $H_s \approx 0.75$ m. It shows that peak pressures may give values between 0.25 and 3 m water column, whereas the width of impact may be between 0.15 and 1 m, with an average value around 0.4 m. But there is hardly any correlation between both parameters.

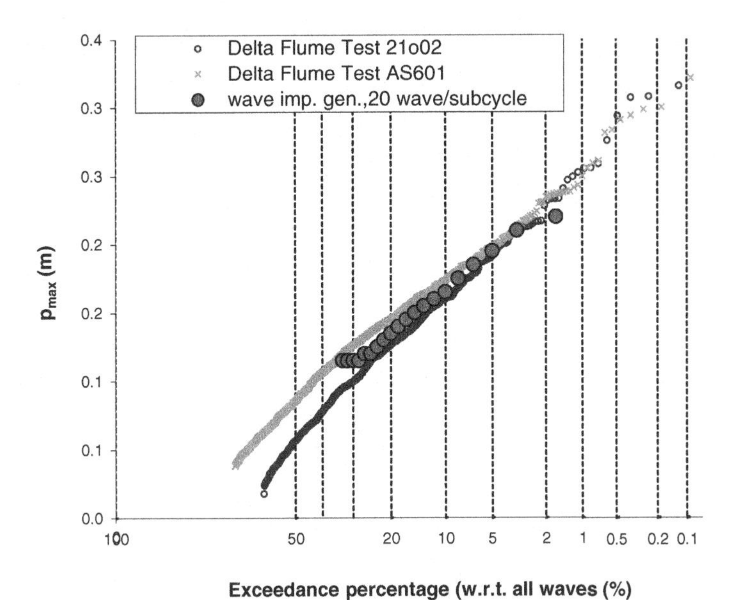

Figure 2. Peak pressures of impacts, measured in the Delta Flume and given on Rayleigh paper. Also simulated pressures are shown (described later in the chapter)

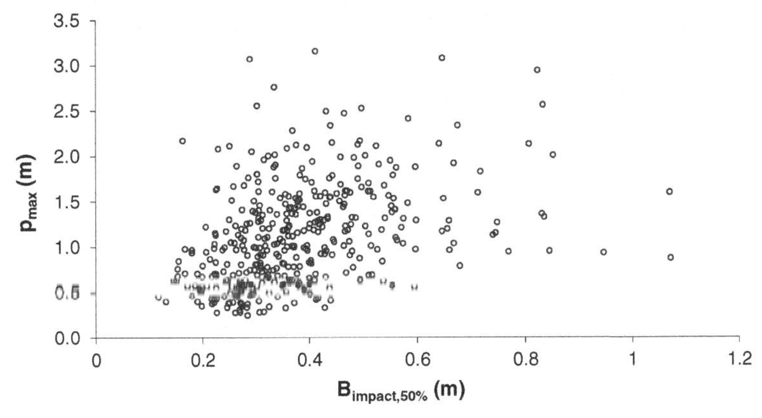

Figure 3. Peak pressures of impacts versus the width of the impacts (Delta flume measurements [Klein Breteler, 2012]).

2.3. *Wave run-up and run-down*

The engineering design parameter for wave run-up is the level on the slope that is exceeded by 2% of the up-rushing waves ($Ru_{2\%}$). The

EurOtop Manual [2007] gives methods to calculate the overtopping discharge as well as the 2% run-up level for all kinds of wave conditions and for many types of coastal structures. Knowing the 2% run-up level for a certain condition is the starting point to describe the wave run-up process. Assuming a Rayleigh distribution of the run-up levels and knowing $Ru_{2\%}$ gives all the required run-up levels. As the EurOtop Manual [2007] is readily available, formulae for wave run-up have not been repeated here.

The wave run-up level is a start, but also run-up velocities and flow thicknesses are required. From the wave overtopping tests it is known that the *front velocity* is the governing parameter in initiating damage to a grassed slope. Focus should therefore be on describing this front velocity along the upper slope. By only considering random waves and the 2%-values, the equations for run-up velocity and flow thickness become:

$$u_{2\%} = c_{u2\%}(g(Ru_{2\%} - z_A))^{0.5} \tag{2}$$

$$h_{2\%} = c_{h2\%}(Ru_{2\%} - z_A) \tag{3}$$

where:

$u_{2\%}$	=	run-up velocity exceeded by 2% of the up-rushing waves
$c_{u2\%}$	=	coefficient
g	=	acceleration of gravity
$Ru_{2\%}$	=	maximum level of wave run-up related to the still water level swl
z_A	=	location on the seaward slope, in the run-up zone, related to swl
$h_{2\%}$	=	flow thickness exceeded by 2% of the up-rushing waves
$c_{h2\%}$	=	coefficient

The main issue is to find the correct values of $c_{u2\%}$ and $c_{h2\%}$. But comparing the results of various research studies [Van der Meer *et al.*, 2012] gives the conclusion that they are not consistent. The best conclusion at this moment is to take $c_{h2\%} = 0.20$ for slopes of 1:3 and 1:4 and $c_{h2\%} = 0.30$ for a slope of 1:6. Consequently, a slope of 1:5 would then by interpolation give $c_{h2\%} = 0.25$. This procedure is better than to use a formula like $c_{h2\%} = 0.055 \cot\alpha$, as given in EurOtop [2007]. One can take $c_{u2\%} = 1.4\text{-}1.5$ for slopes between 1:3 and 1:6.

Moreover, the general form of Equation 2 for the maximum velocity somewhere on a slope, may differ from the front velocity of the up-rushing wave. Van der Meer [2011] analyzed individual waves rushing up the slope. Based on this analysis the following conclusion on the location of maximum or large velocities and front velocities in the run-up of waves on the seaward slope of a smooth dike can be drawn, which is also shown graphically in Figure 4.

In average the run-up starts at a level of 15% of the maximum run-up level, with a front velocity close to the maximum front velocity and this velocity is more or less constant until a level of 75% of the maximum run-up level. The real maximum front velocity in average is reached between 30%-40% of the maximum run-up level. Figure 4 also shows that a square root function as assumed in Eq. 2, which is valid for a maximum velocity at a certain location (not the front velocity) is different from the front velocity. The process of a breaking and impacting wave on the slope has influence on the run-up, it gives a kind of acceleration to the up-rushing water. This is the reason why the front velocity is quite constant over a large part of the run-up area.

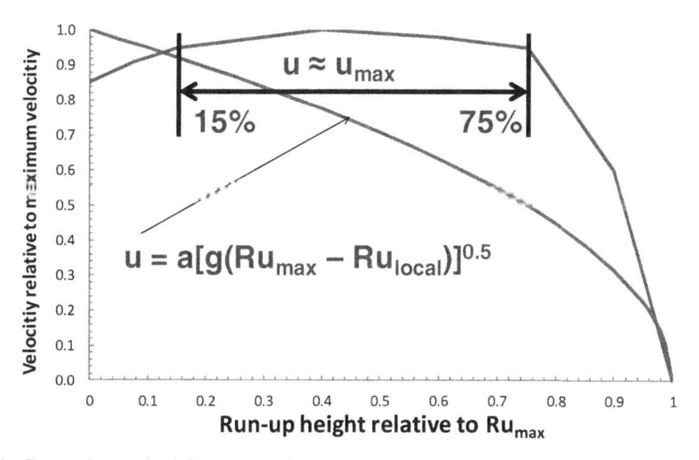

Figure 4. General trend of front velocity over the slope during up-rush, compared to the theoretical maximum velocity at a certain location.

Further analysis showed that there is a clear trend between the maximum front velocity in each up-rushing wave and the (maximum) run-up level itself, although there is considerable scatter. Figure 5 shows the final overall figure (detailed analysis in Van der Meer, [2011]), where front velocity and maximum run-up level of each wave were made dimensionless. Note that only the largest front velocities have been analysed and that the lower left corner of the graph in reality has a lot of data, but less significant with respect to effect on a grassed slope.

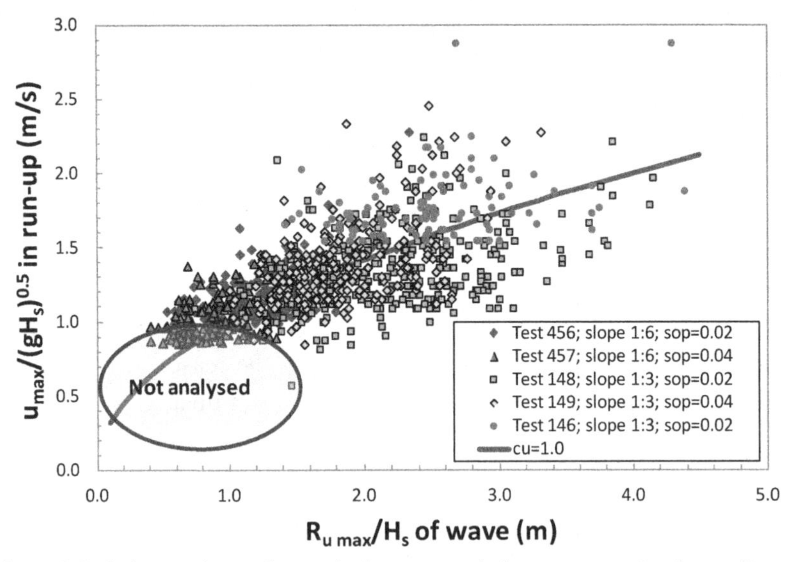

Figure 5. Relative maximum front velocity versus relative run-up on the slope; all tests.

The trend and conclusion in Figure 4 explains for a part why the relationship between the maximum front velocity and the maximum run-up in Figure 5 gives a lot of scatter. A front velocity close to the maximum velocity is present over a large part of the slope and the actual location of the maximum velocity may be more or less "by accident". The trend given in Figure 5 can be described by:

$$u_{max}/\sqrt{(gH_s}=c_u\sqrt{Ru_{max}/H_s} \quad (4)$$

with c_u as stochastic variable ($\mu(c_u)$ = 1.0, a normal distribution with coefficient of variation CoV = 0.25).

2.4. *Wave overtopping*

Like for wave run-up the EurOtop Manual [2007] gives the formulae for also for mean wave overtopping. This is the governing design parameter, which will not be repeated here. In reality there is no mean discharge, but several individual waves overtopping the structure, each with a certain overtopping volume, V. Recent improvements in describing wave overtopping processes have been described by Hughes *et al.,* [2012] and Zanuttigh *et al.*, [2013]. The distribution of individual overtopping wave volumes can well be represented by the two parameter Weibull probability distribution, given by the percent exceedance distribution in Equation 5.

$$P_{V\%} = P(V_i \geq V) = exp\left[-\left(\frac{V}{a}\right)^b\right] \cdot (100\%) \tag{5}$$

where P_V is the probability that an individual wave volume (V_i) will be less than a specified volume (V), and $P_{V\%}$ is the percentage of wave volumes that will exceed the specified volume (V). The two parameters of the Weibull distribution are the non-dimensional shape factor, b, that helps define the extreme tail of the distribution and the dimensional scale factor, a, that normalizes the distribution.

$$a = \left(\frac{1}{\Gamma(1+\frac{1}{b})}\right)\left(\frac{qT_m}{P_{ov}}\right) \tag{6}$$

where Γ is the mathematical gamma function.
Zanuttigh *et al.*, [2012] give for b the following relationship (Fig. 6):

$$b = 0.73 + 55\left(\frac{q}{gH_{m0}T_{m-1,0}}\right)^{0.8} \tag{7}$$

Figure 6 shows that for a relative discharge of $q/(gH_{m0}T_{m-1,0}) = 5.10^{-3}$ the average value of b is about 0.75 and this value has long been used to describe overtopping of individual wave volumes (as given in EurOtop, [2007]). But the graph shows that with larger relative discharge the b-value may increase significantly, leading to a gentler distribution of overtopping wave volumes. This new knowledge may have effect on design and usage of wave overtopping simulators.

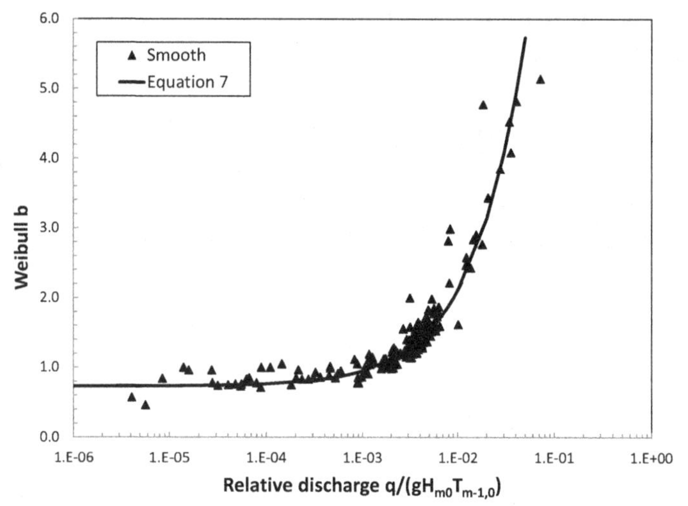

Figure 6. New Weibull shape factor, b, spanning a large range of relative freeboards (Zanuttigh *et al.*, [2013].

3. Simulators as Hydraulic Test Facilities

In total three types of Simulators have been developed, on impacts, run-up and on overtopping. The principle is similar for all three types: a box with a certain geometry is constantly filled with water by a (large) pump. The box is equipped with one or more valves to hold and release the water and has a specifically designed outflow device to guide the water in a correct way to the slope of the dike. By changing the released volume of water from the box one can vary the wave-structure-interaction properties.

3.1. *Wave impacts*

The Wave Impact Generator is a development under the WTI 2017-program of the Dutch Rijkswaterstaat and Deltares, see Figure 7. This tool is called a generator and not a simulator. It has been developed late 2011 and in 2012 and testing has been performed the first and second half of 2012. It is a box of 0.4 m wide, 2 m long and can be up to 2 m high (modular system). It has a very advanced system of two flap valves of only 0.2 m wide, which open in a split second and which enables the water to reach the slope at almost the same moment over the full width of 0.4 m and thus creating a nice impact. Measured impacts are given in Figure 2 and compared with impacts measured in the Delta Flume.

As the location of impacts varies on the slope, the Wave Impact Generator has been attached to a tractor or excavator, which moves the simulator a little up and down the slope. In this way the impacts do not occur all at the same location. Development and description of first tests have been described by Van Steeg [2012a, 2012b and 2013].

Figure 7. Test with Wave Impact Generator.

The main application is simulation of wave impacts on grassed slopes of dikes, like for river dikes, where the wave heights are limited to H_s = 0.5 - 1 m. The impact pressures to be simulated are given by Eq. 1, but within the range of wave heights given here. The impact pressure can be regulated by the empirically determined formula:

$$p_{max} = 1.10h_w + 0.87 \quad (8)$$

where h_w is the water column in the box, with p_{max} measured in m water column. This relation has been calibrated for $0.25 \text{ m} < h_w < 1.0 \text{ m}$. In fact only the largest 30% of the wave impacts is simulated, see also Figure 2.

Slopes with various quality of grass as well as soil (clay and sand) have been tested as well as a number of transitions, which are often found in dikes and which in many cases fail faster than a grassed slope. Figure 8 gives an impression of a road crossing of open tiles, which failed by undermining due to simulated wave impacts.

Figure 8. Failed road crossing by under-mining due to simulated wave impacts.

3.2. *Wave run-up and run-down*

The process of run-up was explored, see Section 2.3, as well as a procedure for testing was developed [Van der Meer, 2011] and [Van der Meer *et al.*, [2012]. Then in 2012 a pilot test was performed on wave run-up simulation, but using the existing Wave Overtopping Simulator as an existing tool (description in the next section). The Simulator was placed on a seaward berm and run-up levels were calibrated with released wave volumes and these were used for steering the process. In this way the largest run-up levels of a hypothetical storm and storm surge, which would reach the upper slope above the seaward berm, were simulated. Figure 9 gives the set-up of the pilot test and shows a wave run-up that even reached the crest, more than 3 m higher than the level of the Simulator.

An example on damage developed by simulating wave run-up is shown in Figure 10. The up-rushing waves meet the upper slope of the dike and "eat" into it.

Figure 9. Set-up of the pilot wave run-up test at Tholen, using the existing Wave Overtopping Simulator.

Figure 10. Final damage after the pilot run-up test.

The pilot test gave valuable information on how testing in future could be improved, but also how a real Wave Run-up Simulator should look like. A Wave Run-up Simulator should have a slender shape, different from the present Wave Overtopping Simulator, in order to release less water, but with a higher velocity. At the end of 2013 such a new device was designed, constructed and tested. And before spring 2014 the first tests on the upper slope of the seaward part of a sea dike was tested. The box had a cross-section at the lower part of 0.4 m by 2 m, giving a test section of 2 m wide, see Figure 11. The upper part had a cross-section of 0.8 m by 1.0 m and this change was designed in order to have less wind forces on the Simulator. The cross-sectional area was the same over the full height of the Simulator in order not to have dissipation of energy during release of water. The overall height is more than 8 m.

Figure 11. The new Wave Run-up Simulator, designed in 2013.

A new type of valve was designed to cope with the very high water pressures (more than 7 m water column). A drawer type valve mechanism was designed with two valves moving horizontally over girders. In this way leakage by high water pressures was diminished as higher pressures gave a higher closing pressure on the valves. This new Wave Run-up Simulator was calibrated against a 1:2.7 slope. The largest run-up was about 13.5 m along the slope, this is about 4.7 m measured vertically. Besides transitions from down slope to berm and berm to upper grassed slope, also a stair case was tested by wave run-up, see Figure 2. As in many other tests, with as well impact or overtopping waves, a stair case is always a weak point in a dike.

Figure 12. Testing a stair case with the new Wave Run-up Simulator in 2014.

3.3. *Wave overtopping*

The Wave Overtopping Simulator has been designed and constructed in 2006 and has been used since then for destructive tests on dike crest and landward slopes of dikes or levees under loading of overtopping waves. References are Van der Meer *et al.*, [2006, 2007, 2008, 2009, 2010, 2011, 2012], Akkerman *et al.*, [2007], Steendam *et al*,. [2008, 2010, 2011] and Hoffmans *et al*,. [2008], including development of Overtopping Simulators in Vietnam [Le Hai Trung *et al*,. 2010] and in the USA [Van der Meer *et al.*, 2011] and [Thornton *et al.*, 2011].

The setup of the Overtopping Simulator on a dike or levee is given in Figure 13, where the Simulator itself has been placed on the seaward slope and it releases the overtopping wave volume on the crest, which is then guided down the landward side of the dike. Water is pumped into a box and released now and then through a butterfly valve, simulating an overtopping wave volume. Electrical and hydraulic power packs enable pumping and opening and closing of the valve. A measuring cabin has been placed close to the test section. The Simulator is 4 m wide and has a maximum capacity of 22 m^2, or 5.5 m^3 per m width. The Simulator in Vietnam has the same capacity, but the Simulator in the US has a capacity of 16 m^3 per m width (although over a width of 1.8 m instead of 4 m). Released volumes in a certain time are according to theoretical distributions of overtopping wave volumes, as described in this chapter, depending on assumed wave conditions at the sea side and assumed crest freeboard.

Figure 13. Set-up of the Wave Overtopping Simulator close to a highway.

Figure 14 shows the release of a large overtopping wave volume and Figure 15 shows one of the many examples of a failed dike section, here a sand dike covered with good quality grass.

Figure 14. Release of a large wave volume.

Figure 15. Failure of a sand dike.

3.4. *Wave impacts by wave overtopping*

One application of a Hydraulic Simulator is to test the resistance of a grass dike by destructive testing, as described in Sections 3.1-3.3. Another application came up different from destructive testing and that is the simulation of wave impacts and measurement of pressures and forces on a structure. The impacts were not generated by wave breaking, like for the Wave Impact Simulator, but by overtopping wave volumes. Two examples will be given here.

The relatively short Belgian coast has a sandy foreshore, protected by a sloping seawall, a promenade and then apartments. In order to increase safety against flooding, vertical storm walls were designed on the promenade. Under design conditions waves would break on the sloping revetment. giving large overtopping waves that travelled some distance over the promenade before hitting the storm wall. Impacts in small scale model investigations may differ significantly from the real situation with larger waves and often salt water (different behaviour of air bubbles compared with fresh water). A full scale test was set-up for the Belgian situation, see Figure 16, with the Wave Overtopping Simulator releasing the flow of overtopping wave volumes over a horizontal distance on to vertical plates where forces as well as pressures could be measured. The tests and results have been described in Van Doorslaer *et al.*, [2012].

Figure 16. Measuring impacts by overtopping waves on a promenade towards a vertical storm wall.

In Den Oever, The Netherlands, a 300 m long dike improvement was designed with the shape of a stair case. The steps were 0.46 m high (sitting height) and 2 m wide and the total structure had four steps. The design wave height was about 1.35 m, which broke over a 6.5 m quay area with the crest at the design water level. The overtopping wave front hit the front side of the stair case type structure, giving very high impacts in a small scale model investigation. The new Wave Run-up Simulator was used to simulate similar impacts, but now on full scale and with salt water. Figure 17 shows the impact of a wave on the lower step of the stair case (the other steps were not modelled).

Figure 17. Measuring impacts by overtopping waves on a quay area towards a vertical step of a stair case type structure.

4. Summary and Discussion

Erosion of grassed slopes by wave attack is not easy to investigate as one has to work at real scale, due to the fact that the strength of clay with grass roots cannot be scaled down. There are two ways to perform tests on real scale: bring (pieces of) the dike to a large scale facility that can produce significant wave heights of at least 1 m, or bring (simulated) wave attack to a real dike. For investigation in a large scale facility the main advantage will be that the waves are generated well and consequently also the wave-structure-interaction processes are generated well. The disadvantage is that the modelled dike has to be taken from a real dike in undisturbed pieces. This is difficult and expensive and real situations on a dike, like staircases, fences and trees are almost impossible to replicate. This type of research is often focussed on the grass cover with under laying clay layer only.

The second alternative of Simulators at a dike has the significant advantage that real and undisturbed situations can be investigated. The research on wave overtopping has already given the main conclusion that *it is not the grass cover itself that will lead to failure of a dike by overtopping*, but an obstacle (tree; pole; staircase) or transition (dike crossing; from slope to toe or berm). The main disadvantage of using Simulators is that only a part of the wave-structure-interaction can be simulated and the quality of this simulation depends on the knowledge of the process to simulate and the capabilities of the device. The experience of testing with the three Simulators, on wave impacts, wave run-up and wave overtopping, gave in only seven years a tremendous increase in knowledge of dike strength and resulted in predictive models for safety assessment or design.

By simulating overtopping waves it is also possible to measure impact pressures and forces on structures that are hit by these overtopping waves. Such a test will be at full scale and if necessary with salt water, giving realistic wave impacts without significant scale or model effects.

Acknowledgments

Development and research was commissioned by a number of clients, such as the Dutch Rijkswaterstaat, Centre for Water Management, the Flemish Government, the USACE and local Water Boards. The research was performed by a consortium of partners and was mostly led by Deltares. Consortium partners were Deltares (project leader - André van Hoven and Paul van Steeg, geotechnical issues, model descriptions, hydraulic measurements, performance of wave impact generator), Infram (Gosse Jan Steendam, logistic operation of testing), Alterra (grass issues), Royal Haskoning (consulting), Van der Meer Consulting (performance of Simulators and hydraulic measurements) and Van der Meer Innovations (Gerben van der Meer, mechanical design of the Simulators).

References

1. Akkerman, G.J., P. Bernardini, J.W. van der Meer, H. Verheij and A. van Hoven (2007). *Field tests on sea defences subject to wave overtopping.* Proc. Coastal Structures, Venice, Italy.
2. EurOtop (2007). *European Manual for the Assessment of Wave Overtopping.* Pullen, T. Allsop, N.W.H. Bruce, T., Kortenhaus, A., Schüttrumpf, H. and Van der Meer, J.W. www.overtopping-manual.com.
3. Hoffmans, G., G.J. Akkerman, H. Verheij, A. van Hoven and J.W. van der Meer (2008). *The erodibility of grassed inner dike slopes against wave overtopping.* ASCE, Proc. ICCE 2008, Hamburg, 3224-3236.
4. Hughes, S, C. Thornton, J.W. van der Meer and B. Scholl (2012). *Improvements in describing wave overtopping processes.* ASCE, Proc. ICCE 2012, Santander, Spain.
5. Klein Breteler, M., van der Werf, I., Wenneker, I., (2012), *Kwantificering golfbelasting en invloed lange golven. In Dutch. (Quantification of wave loads and influence of long waves)*, Deltares report H4421, 1204727, March 2012
6. Le Hai Trung, J.W. van der Meer, G.J. Schiereck, Vu Minh Cath and G. van der Meer. 2010. *Wave Overtopping Simulator Tests in Vietnam.* ASCE, Proc. ICCE 2010, Shanghai.
7. Schüttrumpf, H.F.R. 2001. *Wellenüberlaufströmung bei See-deichen,* Ph.D.-th. Techn. Un. Braunschweig.
8. Steendam, G.J., W. de Vries, J.W. van der Meer, A. van Hoven, G. de Raat and J.Y. Frissel. 2008. *Influence of management and maintenance on erosive impact of wave overtopping on grass covered slopes of dikes; Tests.* Proc. FloodRisk, Oxford, UK. Flood Risk Management: Research and Practice – Samuels et al. (eds.) ISBN 978-0-415-48507-4; pp 523-533.
9. Steendam, G.J., J.W. van der Meer, B. Hardeman and A. van Hoven. 2010. *Destructive wave overtopping tests on grass covered landward slopes of dikes and transitions to berms.* ASCE, Proc. ICCE 2010.
10. Steendam, G.J., P. Peeters , J.W. van der Meer, K. Van Doorslaer, and K. Trouw. 2011. *Destructive wave overtopping tests on Flemish dikes.* ASCE, Proc. Coastal Structures 2011, Yokohama, Japan.
11. Thornton, C., J.W. van der Meer and S.A. Hughes. 2011. *Testing levee slope resiliency at the new Colorado State University Wave Overtopping Test Facility.* Proc. Coastal Structures 2011, Japan.
12. Van der Meer, J.W., P. Bernardini, W. Snijders and H.J. Regeling. 2006. *The wave overtopping simulator.* ASCE, ICCE 2006, San Diego, pp. 4654 - 4666.
13. Van der Meer, J.W., P. Bernardini, G.J. Steendam, G.J. Akkerman and G.J.C.M. Hoffmans. 2007. *The wave overtopping simulator in action.* Proc. Coastal Structures, Venice, Italy.

14. Van der Meer, J.W., G.J. Steendam, G. de Raat and P. Bernardini. 2008. *Further developments on the wave overtopping simulator*. ASCE, Proc. ICCE 2008, Hamburg, 2957-2969.
15. Van der Meer, J.W., R. Schrijver, B. Hardeman, A. van Hoven, H. Verheij and G.J. Steendam. 2009. *Guidance on erosion resistance of inner slopes of dikes from three years of testing with the Wave Overtopping Simulator*. Proc. ICE, Coasts, Marine Structures and Breakwaters 2009, Edinburgh, UK.
16. Van der Meer, J.W., B. Hardeman, G.J. Steendam, H. Schttrumpf and H. Verheij. 2010. *Flow depths and velocities at crest and inner slope of a dike, in theory and with the Wave Overtopping Simulator*. ASCE, Proc. ICCE 2010, Shanghai.
17. Van der Meer, J.W., C. Thornton and S. Hughes. 2011. *Design and operation of the US Wave Overtopping Simulator*. ASCE, Proc. Coastal Structures 2011, Yokohama, Japan.
18. Van der Meer, J.W. 2011. *The Wave Run-up Simulator. Idea, necessity, theoretical background and design*. Van der Meer Consulting Report vdm11355.
19. Van der Meer, J.W., Y. Provoost and G.J. Steendam (2012). *The wave run-up simulator, theory and first pilot test*. ASCE, Proc. ICCE 2012, Santander, Spain.
20. Van Doorslaer, K., J. De Rouck, K. Trouw, J.W. van der Meer and S. Schimmels. *Wave forces on storm walls, small and large scale experiments*. Proc. COPEDEC 2012, Chennai, India
21. Van Steeg, P., (2012a). *Reststerkte van gras op rivierdijken bij golfbelasting. SBW onderzoek. Fase 1a: Ontwikkeling golfklapgenerator. In Dutch (Residual strength of grass on river dikes under wave attack. SBW research Phase 1a: Development wave impact generator)*. Deltares report 1206012-012-012-HYE-0002, April 2012
22. Van Steeg, P., (2012b). *Residual strength of grass on river dikes under wave attack. SBW research Phase 1b: Development of improved wave impact generator*. Deltares report 1206012-012-012-HYE-0015 August 2012
23. Van Steeg, P., (2013) *Residual strength of grass on river dikes under wave attack. SBW research Phase 1c: Evaluation of wave impact generator and measurements based on prototype testing on the Sedyk near Oosterbierum*. Deltares report 1207811-008.
24. Zanuttigh, B., J.W. van der Meer and T. Bruce, 2013. *Statistical characterisation of extreme overtopping wave volumes*. Proc. ICE, Coasts, Marine Structures and Breakwaters 2013, Edinburgh, UK.

CHAPTER 2

DESIGN, CONSTRUCTION AND PERFORMANCE OF THE MAIN BREAKWATER OF THE NEW OUTER PORT AT PUNTO LANGOSTEIRA, LA CORUNA, SPAIN

Hans F. Burcharth

Professor, Port and Coastal Engineering Consultant. Aalborg University, Denmark.
E-mail: hansburcharth@gmail.com

Enrique Maciñeira Alonso

Port Planning and Strategy Manager, Port Authority of La Coruña, Spain.
E-mail: emacine@puertocoruna.com

Fernando Noya Arquero

Port Infrastructure Manager, Port Authority of La Coruña, Spain.
E-mail: fnoya@puertocoruna.com

1. Introduction

The port of La Coruna at the North-West shoulder of Spain is situated in the inner part of a bay surrounded by the city, see Fig. 1. Besides general cargo the port is handling hydrocarbons, coal and other environmentally dangerous materials. Recent accidents made it clear that the close proximity to the town was not acceptable and in 1992 it was decided to look for a new location for the potentially dangerous port activities. Many locations were investigated considering land and sea logistics, economy and environmental risks. The final choice, decided in 1997, was a location at Cape Langosteira on the open coast facing the North

Atlantic Ocean, approximately 9 km. from the city but closer to the Repsol petrochemical refinery. A very large breakwater was needed for the protection of the port. Design significant wave height exceeding 15 m, water depths up to 40 m, and a very harsh wave climate made it a challenge to design and construct the 3.4 km long breakwater. Fig. 1 shows an air photo of the area covering the old port of La Coruna as well as the new port at Punta Langosteira.

Fig. 1. The old port in the center of La Coruna city, and the new port at Punta Langosteira under construction.

A minimum water depth of 22 m was needed in the port for the liquid bulk carriers and for the container and general cargo vessels. The final layout of the port is shown in Fig. 2. The liquid bulk berths are to be arranged perpendicular to the breakwater.

The present article deals with the environmental conditions, the design, the model testing and the construction of the main breakwater including reliability analysis and experienced performance of the structure.

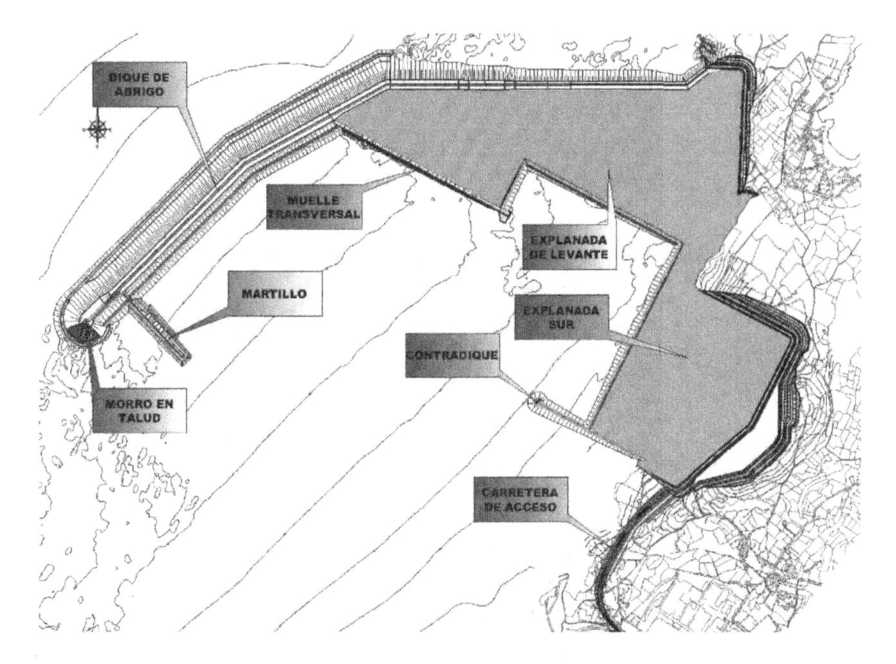

Fig. 2. Final layout of the port.

2. Environmental Conditions

2.1. *Water levels*

LLWS = 0. 00 m. HHWS = + 4.50 m. The maximum storm surge can be estimated to app. 0.40 – 0.50 m.

The design level was set to + 4.5 m.

2.2. *Bathymetry and seabed*

The bathymetry of the area is shown in Fig. 3. A shoal with top at level - 30 m is seen at some distance from the coast. The refraction and diffraction of the waves caused by the shoal were studied in a large physical model at Delft Hydraulics and in numerical models, [1]. Short crested waves (directional waves) were used in both cases. The investigations showed that the shoal causes a rather significant increase of 13% in wave heights along the outer part of the breakwater.

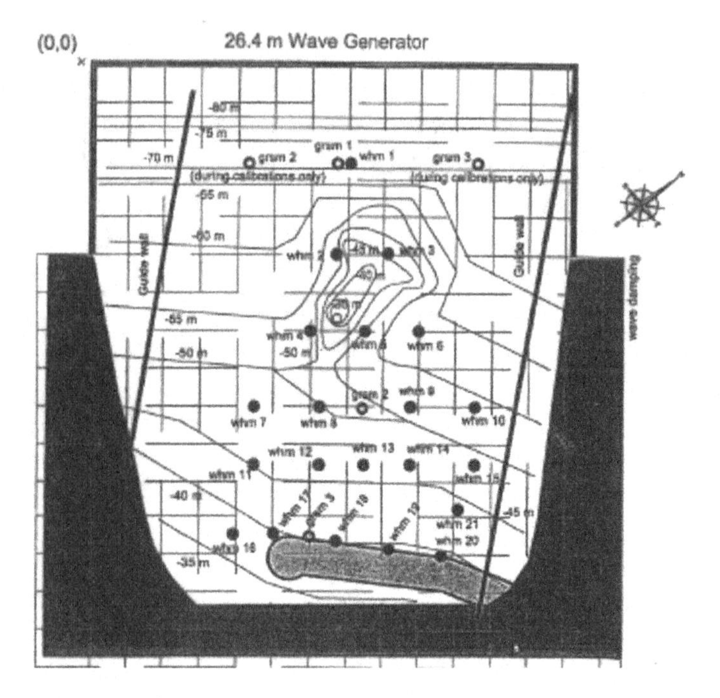

Fig. 3. Bathymetry of the model in the multidirectional wave basin at Delft Hydraulics for testing of the influence of the shoal on the waves along the breakwater.

The seabed consists of rock which is only spot wise covered with thin layers of sand.

2.3. *Waves*

Waves from the directional sectors WNW, NW and NNW are the most critical for the breakwater. The largest waves are from NW having significant wave heights approximately 30-40% larger than from the two other sectors. The NW waves approach perpendicular to the outer part of the breakwater.

The wave data available for the conceptual design of the breakwater in the beginning of the year 2000 was the deep water WASA data (1970 – 1994) derived from a prediction numerical model based on wind data. The data were provided by the Maritime Climate Department of Puertos del Estado in Spain. For prediction of the long-term wave statistics were chosen 54 storms with significant wave heights exceeding 8.5 m covering a 25 years period. Each storm is characterized by a significant wave height H_s, a mean period T_z and a wave direction.

The storms were propagated using the numerical model GHOST from deep water to zones along the breakwater. The data for the outer part of the breakwater in 40 m water depth were fitted using a POT analysis (Peak Over Threshold) to Weibull distributions.

Three methods of fitting the data to the distribution were investigated. The statistical parameters in a 3-parameter Weibull distribution were estimated by the Maximum Likelihood and the Least Square methods, and the two statistical parameters in a truncated Weibull distribution with threshold level 8.00 m were estimated by the Maximum Likelihood method. The best goodness of fit was obtained by the last method. The related return period H_s-values with and without statistical uncertainty are given in the table in Fig. 4 which shows the fitted distribution.

$$F_{H_s^*}(h) = \left(1 - \frac{1}{P_0}\exp\left(-\left(\frac{h}{\beta}\right)^{\alpha}\right)\right)^{\lambda} \quad , \quad h \geq \gamma$$

$$P_0 = \exp\left(-\left(\frac{\gamma}{\beta}\right)^{\alpha}\right)$$

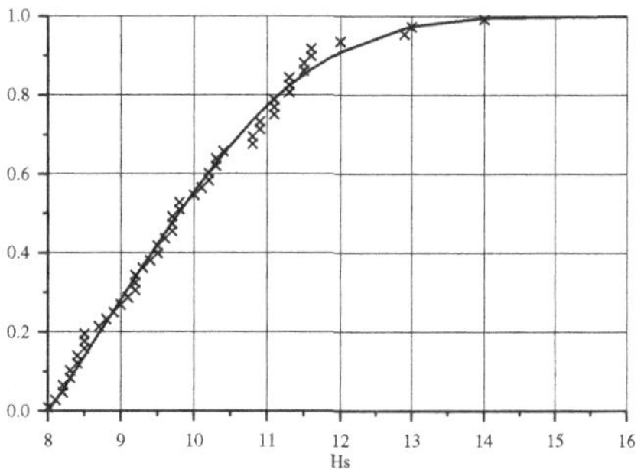

Parameter	γ	α	β
Estimate	8.0 m	4.39	9.43 m
COV		0.24	0.072

Return period [year]	Significant wave height [m] central estimate	Significant wave height [m] incl. statistical uncertainty
25	13.26	13.76
50	13.70	14.45
100	14.10	15.15
140	14.29	15.66
475	14.88	18.77

Fig. 4. Fitting of truncated Weibull distribution and related return period values of H_s for the outer part of the breakwater in 40 m water depth, [2].

The peak periods of the storm waves are in the range T_p = 16 -20 s. The design basis for the initial conceptual design of the most exposed part of the breakwater was set to H_s = 15.0 m, T_p = 20 s.

In 1998 a directional wave recorder named Boya Langosteira was placed approximately 600 m from the outer part of the position of the planned breakwater. Later on at the start of the works another recorder, Boya UTE, was placed in a distance of approximately 400 m to the north.

In the beginning of 2011 almost 14 years of wave data was available from the Boya Langosteira. The data series was expanded by use of the 44 years of SIMAR hindcasted data, [3] from SIMAR 1043074 at position 54 km west of Punta Langosteira. Three years of data (1998-2001) from the very nearby Villano-Sisargas directional buoy (placed in 410 m water depth) were used to calibrate the 44 years of data. The data were then numerically propagated to the position of Boya Langosteira and correlated to the Boya Langosteira data from the same period. On this basis a 44 years data series at the position of Boya Langosteira were established. From this the return period H_s- values for the outer part of the breakwater given in Table 1 were estimated.

Table 1. Return period estimates of H_s at the position of Boya Langosteira, as predicted by transformation and calibration of 44 years of SIMAR data, [3].

Return period (years)	H_s (m). Waves from NW sector
25	11.7
50	12.1
100	12.5
200	12.8
500	13.7

The wave heights given in Table 1 are significantly smaller than those given in Fig. 4, which gave some concern. The reason for this deviation is most probably the rather poor correlation between the SIMAR data and the Villano Sisargas buoy data representing only three years overlap, so finally the 14 years of Boya Langosteira data were preferred for the final prediction of the wave climate. The related wave heights used for the design of the breakwater are given in Table 2.

Table 2. Return period significant H_s –values used for the design of the breakwater, [4].

Return period (years)	H_s (m). Waves from NW sector
25	12.6 – 13.7
50	13.1 – 14.3
100	13.6 – 14.9
140	13.8 – 15.1
475	14.5 – 16.1

Figure 5 shows the distribution of the 140 years return period design waves along the breakwater.

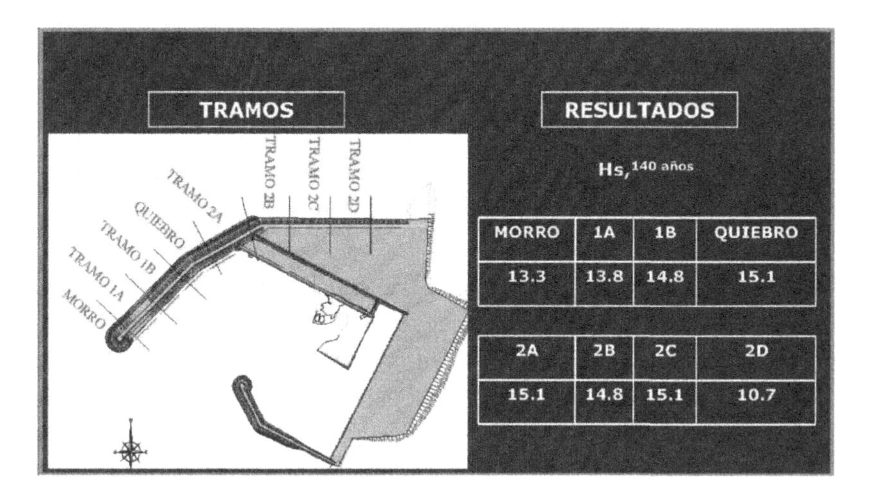

Fig. 5. Distribution of the 140 years return period significant wave heights along the main breakwater.

3. Performance Criteria and Related Safety Level for the Breakwater

The ISO 21650 standard, Actions from Waves and Currents on coastal Structures, specifies that both Serviceability Limit State (SLS) and Ultimate Limit State (ULS) must be considered in the design, and performance criteria and related probability of occurrence in structure service lifetime must be assigned to the limit states.

The Spanish Recommendation for Marine Structures ROM 0.2.90 [5] was used as a basis for the design of the breakwater. According to this the breakwater belongs to Safety Level 2 valid for works and installations of general interest and related moderate risk of loss of human life or environmental damage in case of failure. The assigned structure design lifetime is T_L = 50 years. The Economic Repercussion Index ERI, defined as (costs of losses)/ (investment), is average for the actual structure. The related design safety levels in terms of maximum probability of failure P_f within T_L are:

Initiation of damage, $P_f = 0.30$ (more restrictive than corresponding to SLS)

Total destruction, $P_f = 0.15$ (more severe than corresponding to ULS)

If no parameter and model uncertainties are considered, i.e. only the encounter probability related to storm occurrences is considered then a $P_f = 0.30$ in 50 years corresponds to a return period of 140 years for the design storm. A $P_f = 0.15$ corresponds to a design storm return period of 300 years.

The target damage levels for SLS must correspond to limited damage, e.g. for rock and cube armour layers a maximum of approximately 5% of the unit displaced. The target damage level for ULS corresponds normally to very serious damage without complete destruction, e.g. for rock and cube armour layers approximately 15–20% of the units displaced. The relative number of displaced units in toe berms can be larger as long as the toe still supports the main armour.

Geotechnical slip failures of the concrete super structure are regarded ULS – damage.

The later edition of the Spanish Recommendations for Maritime Structures ROM 0.0 (2002), [6] makes use of a Social and environmental repercussion index (SERI) besides the Economic repercussion index (ERI) as the basis for classification of the structures and related safety levels. Maritime structures are classified in four groups according to little, low, high and very high social and environmental impact.

The Economic Repercussion Index (ERI) is given in Table 3.

Table 3. Economic repercussion values and related minimum design working life, ROM 0.0 (2002).

low:	ERI < 5,	15 years
moderate:	5 < ERI ≤ 20,	25 years
high:	ERI > 20,	50 years

The Social and Environmental Repercussion Index (SERI) and the related maximum probability of failure P_f (and β-values) within the working life as given in Table 3 are presented in Table 4 both for serviceability and ultimate limit states. The failure probability P_f is related to the reliability index β by $P_f = \Phi(-\beta)$ where Φ is the distribution function of a standardized normal distributed stochastic variable.

Table 4. Social an Environmental Repercussion Index (SERI) and related maximum P_f (and β-values) within working life. ROM 0.0 (2002), [6].

Social and environmental impact	Serviceability limit state, SLS		Ultimate limit state, ULS	
	P_f	β	P_f	β
no, SERI < 5	0.20	0.84	0.20	0.84
low, 5 ≤ SERI < 20	0.10	1.28	0.10	1.28
high, 20 ≤ SERI < 30	0.07	1.50	0.01	2.32
very high, SERI ≥ 0	0.07	1.50	0.0001	3.71

As for the actual very large breakwater the ERI is high, and the design working life therefore set to 50 years. SERI is estimated close to 20 for which reason P_f is set to 0.10 both for SLS and ULS.

If no parameter and model uncertainties are considered, i.e. only the encounter probability related to storm occurrences is considered, then a $P_f = 0.10$ in 50 years corresponds to a return period of 500 years for the design storm.

The actual breakwater is however designed in accordance with ROM 0.2.90, which is less restrictive than ROM 0.0, see above, but in better agreement with the optimum safety levels found by the PIANC Working Group 47 for conventional outer breakwaters.

However, in the ROM 1.0.09, [7] published in 2010 safety level recommendations are given for conditions of dangerous goods being moved on and behind the breakwater. For the Punta Langosteira breakwater the safety levels related to SLS and ULS must correspond to P_f = 0.01 and 0.07 respectively in 50 year service lifetime. The corresponding encounter probability design return period is up to app. 5.000 years, in this case corresponding to H_s = 16.5 m. This is the maximum significant wave height used in the model tests in which some damage but no collapse of the structure was observed.

4. Initial Conceptual Design and Model Testing of Trunk Section in 40 m Water Depth

4.1. *Cross section and target sea states*

Figure 6 shows the initial conceptual design as presented by ALATEC Ingenieros Consultores y Arquitectos on the basis of design significant wave height of 15.0 m and a spectral peakperiod of 20 s. The stability of the 150 t concrete cube main armour, the toe berm and the rear slope was tested in medium scale model tests at the Hydraulics and Coastal Engineering Laboratory, Aalborg University, Denmark. The wave forces on the concrete wave wall as well as the average overtopping discharge and its spatial distribution were recorded as well.

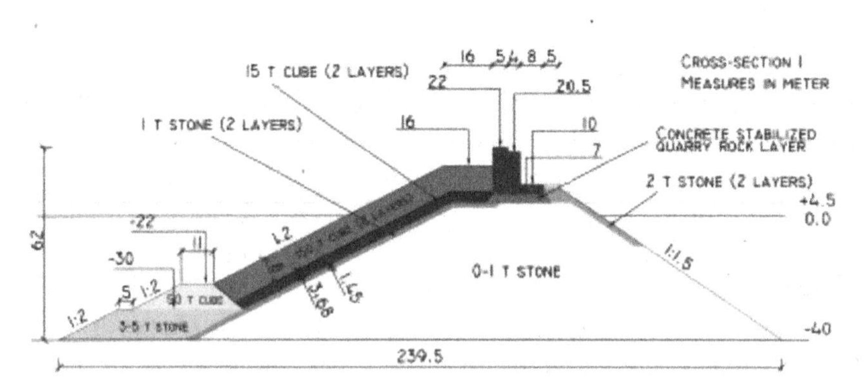

Fig. 6. Initial conceptual cross section tested in a 2–D model at the Hydraulics and Coastal Engineering Laboratory, Aalborg University, Denmark

The 150 t cubes have the dimensions 4 x 4 x 4 m and mass density 2.4 t/m^3.

The length scale of the model tests was 1:80. JONSWAP spectra with peak enhancement factor 3.3 were used in the tests. Significant wave heights in the test series were up to 21 cm. Active absorption of the reflected waves used in all tests was a necessity because the wave reflection form the structure was very large (reflection coefficients in the range 22%–35%) due to the long waves. The core material size was enlarged in accordance with Burcharth et al. (1999), [8] in order to compensate for viscous scale effect. Damage in terms of relative number of displaced armour units was identified by photo overlay technique.

While the stability of the main armour and the toe berm were satisfactory, the stability of the rear slope was not satisfactory. Moreover, the wave forces on the wall were so large that stability of the structure could not be obtained. Also the overtopping was excessive. On the basis of these results eight different cross sections were designed, in which the levels of the main armour crest berm and the crest level of the concrete wall were varied (mainly increased). Also the rear slope armour was changed to larger rocks as well as to pattern placed concrete blocks.

The finally tested cross section in the initial stage of design is shown in Fig. 7. In order to reduce the wave induced forces on the concrete wave wall to a level in which stability could be obtained, it was necessary to raise the armour crest level to + 25.0 m. The wall crest was raised to the same level in order to reduce the overtopping to acceptable levels during normal operation of the port. In storms the overtopping was too large for any traffic on the breakwater.

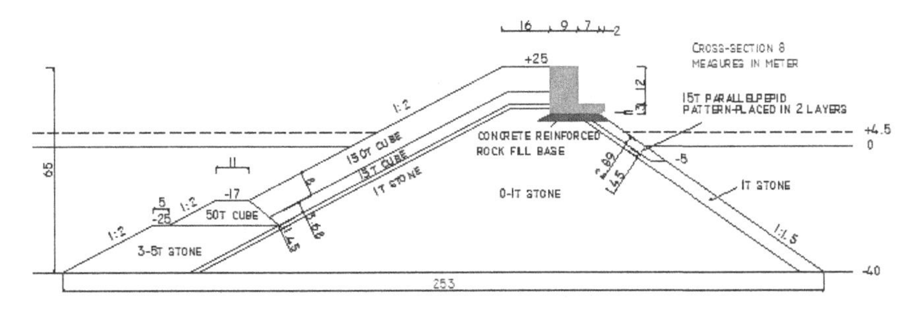

Fig. 7. Finally tested cross section in 2-D tests at Aalborg University

All tests series were repeated several times in order to identify the scatter in the results which was needed for the subsequent reliability analysis of the structure components. Each cross-section model was exposed to a sea state test series consisting of six steps as shown in Table 5, starting with moderate waves and terminating with waves corresponding to an extremely rare and severe storm with probability of occurrence well below that of the design storm.

Table 5. Target sea states in a test series

Hs (m)	8	10	12.5	14.5	15	16.5
Tp (s)	15	15	18	20	20	20

Before the start of a test series, the slope was rebuilt, the pressure transducers and wave gauges were calibrated, and photos were taken by a digital camera.

Each sea state had a duration of app. 5.2 hours (35 minutes in the model), giving 1080-1200 individual waves. After expose of the model to each sea state, photos were taken of the armour layer, the toe-berm and the rear slope.

4.2. *Main armour damage development with increasing wave height*

The typical development of the main armour layer damage with increasing wave height is shown in Fig. 8. In general, the results fit well with the formula of Van derMeer (1988), [9] (modified from validity slope 1:1.5 to slope 1:2, cf. Paragraph 10.3.1) as shown in the figure.

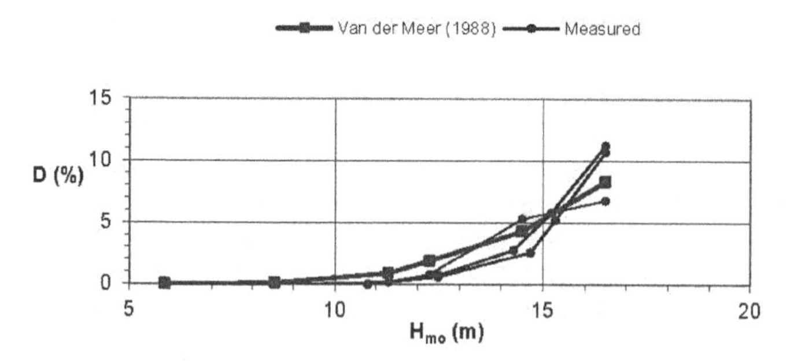

Fig. 8. An example of the main armour layer damage development with increasing wave height. Each sea state contains a little more than 1000 individual waves. Results from repeated tests of the cross section shown in Fig. 6 with the water level +4.5m.

Main armour damage development with storm duration

The development of the main armour layer damage with the storm duration is given in Fig. 9.

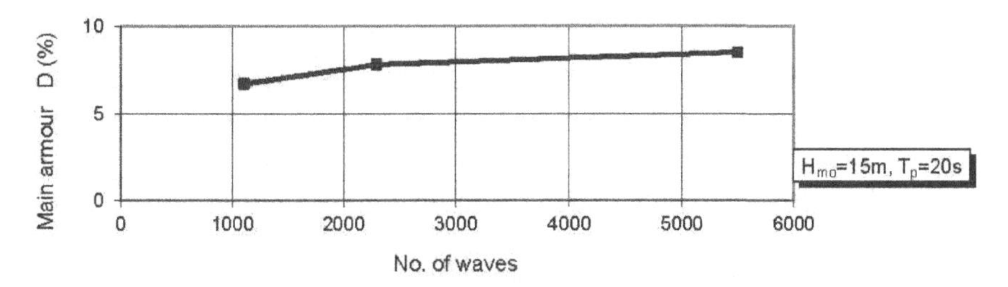

Fig. 9. Example of main armour damage development with storm duration.

An example photo of the slope after wave attack is presented in Fig. 10.

Fig. 10. Slope after Hs=15.2m, Tp=19.2s. From the test series of the cross-section shown in Fig. 7 with the water level +4.5m.

It can be seen that the filter layer is still protected by the 150t cubes after exposure to storm waves slightly larger than the design wave.

The observed damage level in terms of relative number of displaced cubes was for the design sea state Hs = 15.0m, Tp = 20 s in the range D = 3.9 - 8% with an average of 5.4%. This corresponds very well to what is acceptable for a SLS.

The cross section was tested in waves with up to Hs = 16.5m. This corresponds to a return period of approximately 500 years or more. The damage to the main armour was still below D = 15%, i.e. quite far from total destruction.

The Hs values 15.0 m and 16.5 m correspond for Δ = 1.34 to the stability factors KD = 11 and 14 in the Hudson equation. The related Ns = Hs/ (ΔDn) are 2.8 and 3.1, respectively. The corresponding stability factors in the Van der Meer cube armour formula [8], (modified from slope 1:1.5, which corresponds to the validity of the formula, to slope 1:2) are Nod = 0.85 and 1.48, respectively. The formula is given in Paragraph 10.3.1.

4.3. *Toe stability*

The higher the toe, the less reach needed for the crane placing the main armour. The highest acceptable toe level was found to be –17 m. For this the relative displacement of the 50 t cubes in the toe was for the design conditions in the range D = 4.8 – 6.6% with an average of 5.3%, i.e. far from losing its support of the main armour. Moreover, for sea states more severe than the design sea state the toe armour was somewhat protected by displaced 150 t main armour cubes, cf. Fig. 10.

4.4. *Overtopping and rear side armour stability*

The target overtopping discharge was set to 0.150 m^3/(m s) for the design sea state. This implies that no traffic is allowed during storms.

The overtopping discharge and its spatial distribution were measured by collecting the water in trays as shown in Fig. 11.

The overtopping discharge was as expected rather sensitive to the geometry of the cross section and the water level. For two repeated tests with the cross section shown in Fig. 6 the discharges were 0.137 and

0.174 m3/(ms) for water level + 4.50 m, and 0.084 m3/(ms) for water level 0.00 m. The measured overtopping discharges compared very well to predictions by the Pedersen 1996 formula, [10].

In design storms the overtopping water passed the crest of the breakwater and splashed down in the water without hitting the top of the rear slope armour, see Fig. 12. This was a result of the relatively narrow crest, the design of which was shaped to a one-off type of crane running on rails on the concrete base plate as shown in Fig. 13.

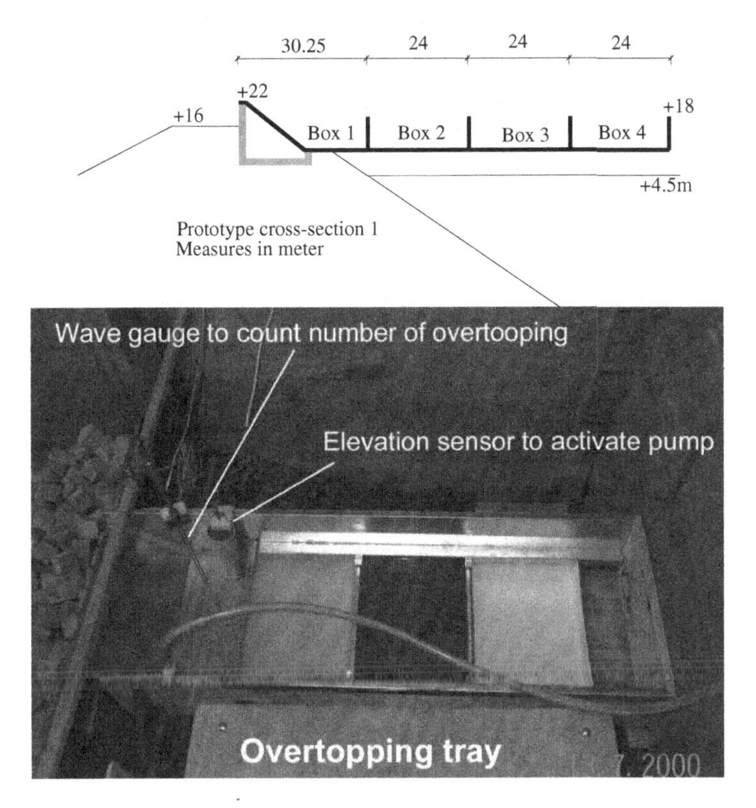

Fig. 11. Arrangement of overtopping trays in the model

Because of much better mobility and flexibility in construction it was later decided to use specially designed crawler cranes. These cranes demanded much more space for which reason the crest width of the breakwater was increased significantly. As a result the rear slope armour was completely redesigned; cf. the discussion later in the article.

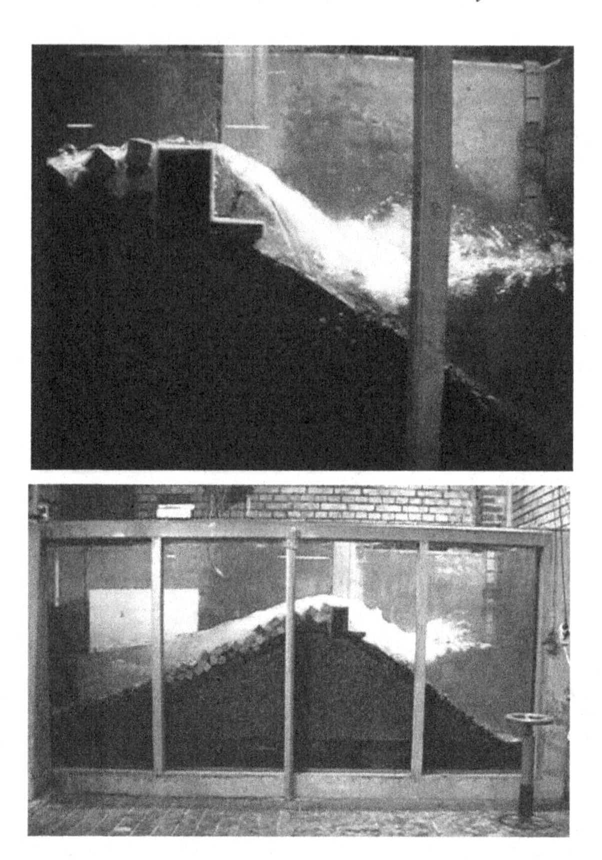

Fig. 12. Typical overtopping splash-down in design sea state

Fig. 13. Initial concept of crane for placement of the armour units

Several solutions for rear side armour were tested. The final solution was a double layer of pattern placed 15 t parallelepiped concrete blocks. The smooth surface makes the armour less vulnerable to forces from the splash down.

Figure 14 shows photos of the rear slope performance in tests with significant wave heights up to H_s = 16.5 m.

Fig. 14. Photos showing rear slope damage development with increasing wave height from tests of the cross section shown in Fig. 7. Water level is +4.50 m. The upper right corner of the rear slope is formed of 1t stone due to lack of 15t concrete parallelepipeds. Hence the damage in the upper right corner should be disregarded.

4.5. *Wave forces on the crown wall and superstructure stability*

A very detailed investigation of the wave induced forces on the concrete superstructure were performed including analyses of the most critical simultaneous combination of horizontal and uplift forces with respect to foundation slip failures. Figure 15 shows the pressure gauges mounted on the superstructure.

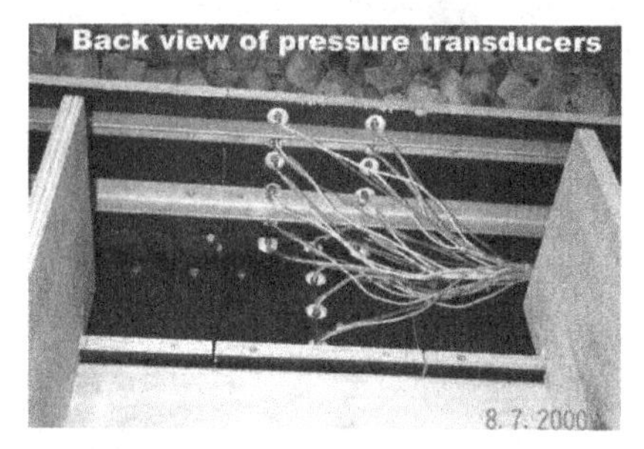

Fig. 15. Pressure gauges for recording of wave induced pressures on superstructure.

In Fig. 16 are shown the special arrangement for measuring the wave induced forces on main armour cubes which, when being in contact with the wall, transferred extra loads to the superstructure.

Fig. 16. Arrangement for measuring wave induced forces transferred to the wave wall by the main armour cubes. The surrounding armour blocks not yet placed.

Figure 17 illustrates the various loads on the superstructure

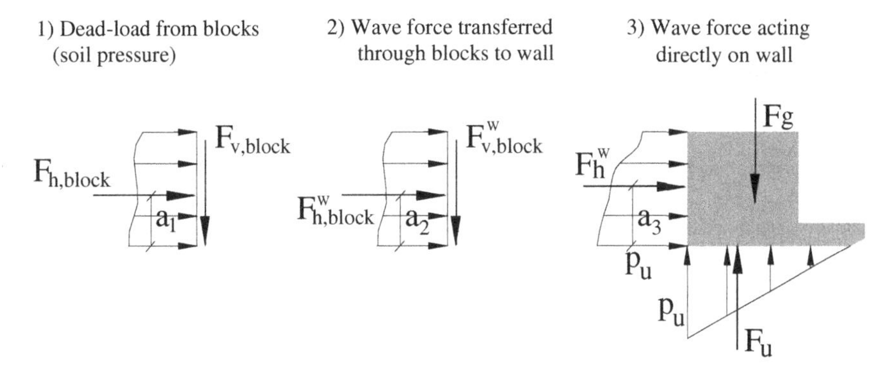

Fig. 17. Illustration of forces on the superstructure

The sampling frequency of the wave induced pressures was 250 Hz. With test series consisting of 1000-1200 waves the maximum loads correspond approximately to the 0.1% exceedance values. Because the maximum values of the horizontal force and the uplift force do not occur simultaneously, it is on the safe side to use the uncorrelated maximum loads in the stability calculations. However, stability calculations were performed both with uncorrelated and correlated forces, the latter corresponding to the most critical combinations of recorded loads.

The failure modes sliding and geotechnical slip failure were analyzed in deterministic stability calculations performed after the tests. The analyses did show that the width of the superstructure should be extended by 2 m from the 18 m shown in Fig. 7 to 20 m, of which the thickness of the wall is 10 m. The following analyses relate to this cross section.

The failure function for *sliding* reads

$$G = f(F_g - F_u) - S F_h, \qquad \text{Failure if } G < 0, \qquad \text{Safe if } G > 0$$

The friction factor $f = 0.7$. The safety factor $S = 1.2$.

The average of the uncorrelated wave forces, i.e. the maximum horizontal wave force $F^W_{h,0.1\%}$ and the simultaneous moment M_h, and the maximum uplift pressure on the front foot of the superstructure $p_{u,0.1\%}$ in all the tests, are for $H_s = 15.00$ m:

$$F^W_{h,0.1\%} = 1215 \text{ kN/m}, \quad a = \frac{M_h}{F^W_{h,0.1\%}} = 5.79 \text{ m}, \quad p_{u,0.1\%} = 120 \text{ kN/m}^2$$

The related $G = 621$ kN/m while for correlated forces $G = 745$ kN/m.

Both values signify a large safety level against sliding.

The failure mode and the related failure function for geotechnical slip failure is shown in Fig. 18.

$$\text{Minimum of } G(\theta) \begin{cases} <0 & \text{Failure} \\ =0 & \text{Limit} \\ >0 & \text{Safe} \end{cases} \qquad G(\theta) = (\rho - \rho_w)\, g\, A\, \omega_V + (F_g - F_u)\, \omega_V - S\,(F_h + F_{hu})\, \omega_H$$

$$\text{for} \quad 0^\circ \leq \theta \leq \tan^{-1}\left(\frac{h_{II}}{B_z + m h_{II}}\right)$$

Horizontal and vertical displacement ω_h and ω_v are function of the effective friction angle φ_d of core material

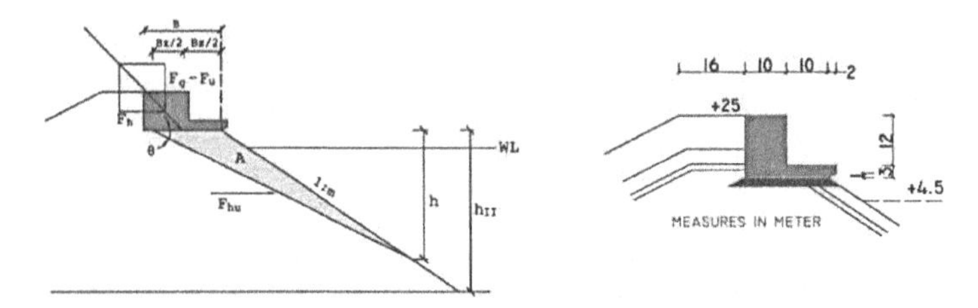

Fig. 18. Geotechnical slip failure mode and related failure function. ρ and ρ_w are mass density of core material and water, respectively.

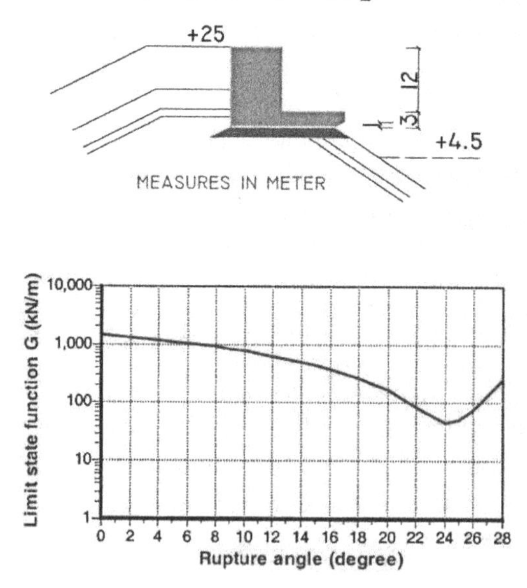

Fig. 19. Minimum values of the failure function for the geotechnical slip failure as function of the rupture angle.

From the recorded wave induced pressures on the superstructure the correlated loadings, which gives the minimum values of the failure function G (θ) were identified. The results are given in Fig. 19 for safety factor S = 1.2 and an adopted core material friction angle of $\varphi_d = 40^0$. It is seen that the most critical rupture angle is $\theta = 24^0$ and that the related minimum value of G (θ) = app. 60 kN/m which indicates an acceptable safety level against slip failures.

5. Probabilistic Safety Analysis of the Final Model Tested Cross Section at Aalborg University

5.1. *Introduction*

A detailed reliability analysis (Sorensen et al. 2000), [2] was performed in order to identify the safety levels of the various parts of the structure as a basis for a more detailed design. The reliability analyses demanded that the uncertainties and scatter in the model test results were documented. For this reason the model tests described above were repeated several times in order to obtain reliable expectation values and coefficients of variation of the parameters.

5.2. *Stochastic models*

5.2.1. *Model for significant wave height*

The truncated Weibull distribution with the statistical parameters given in Fig. 3 was the preferred model for the analysis although other models were applied as well.

5.2.2. *Model for maximum significant wave height*

The model uncertainty related to the quality of the measured wave data is modeled by a multiplicative stochastic variable FH_s, assumed to be normal distributed with expected value 1 and standard deviation 0.1 corresponding to hindcasted wave data (on the safe side). The resulting significant wave height becomes

$$H_s = F_{Hs} H_s^T$$

5.2.3. *Model for tidal elevation*

The tidal elevation ζ is modeled as a stochastic variable with distribution function

$$F_\zeta(\zeta) = 1/\pi \; arcos\,(1 - \zeta/\zeta_o)$$

such that ζ varies between 0 and $2\zeta_o$.

5.2.4. *Modelling of forces on super structure*

Three sets of wave induced horizontal forces, associated moments and uplift pressures were estimated on the basis of measurements performed in the laboratory. The maximum forces are identified using the deterministic limit state function for geotechnical failure in the rubble mound. The following sets of wave induced forces were analyzed:

Set 1. Maximum horizontal forces and corresponding moments and simultaneous uplift pressure
Set 2. Maximum uplift pressure and simultaneous horizontal forces and corresponding moments
Set 3. Simultaneous horizontal forces, moments and uplift pressures resulting in maximum load effect.

The horizontal wave force F_{HW}, the corresponding moment M_{HW} *and* the maximum uplift pressure p_u *at* the front edge of the base plate are assumed to be modelled by a piecewise linear function of H_s. Table 6 gives the expected value (denoted by E [-]) and the standard deviation (denoted by σ [-]) for the forces, moments and uplift pressures at specific significant wave heights for the Set 3 mentioned above.

For each H_s-value the forces, moments and uplift pressures are assumed to be independent and Normal distributed. For given H_s the horizontal force and the uplift force are fully correlated (simultaneous recordings from the model tests).

Table 6. Expected values and standard deviations for Set 3

H_s (m)	$E[F_{HW}]$ (kN/m)	$\sigma[F_{HW}]$ (kN/m)	$E[M_{HW}]$ (kNm/m)	$\sigma[M_{HW}]$ (kNm/m)	$E[p_u]$ (kN/m^3)	$\sigma[p_u]$ (kN/m^3)
11	525	152	2725	325	11	7
12	900	85	4300	585	23	15
13	1033	77	5300	48	46	20
14	1075	73	5950	459	77	29
15	1165	104	7025	971	104	10
16	1330	100	8350	1000	115	10

The measured maximum values of the wave loadings correspond very well to predictions calculated by the Pedersen formula, [10]. The values given in Table 6 correspond to a tidal elevation of $2\zeta_0$. The effect of *varying tidal height* is included by introducing a multiplication factor which reduces the wave induced forces and moments:

$$R_{tide} = 0.74 + 0.26\, U$$

where U is a stochastic variable uniformly distributed between 0 and 1.

The *vertical wave induced uplift force* on the base plate is, on the safe side, assumed triangularly distributed, cf. Fig. 17.

The *horizontal pore pressure force in the core*, acting on the rupture line (see Fig. 17) is estimated by

$F_{HU} = 0.5\, B\, p_U \tan\theta$, in which B is the width of the base plate

The *horizontal dead-load force on the front wall* from the armour blocks resting against the wall is estimated on the basis of a triangular pressure distribution to be

$F_{HD} = 0.5(1 - n)(\rho_C - \rho_W)(1 - \sin\varphi)$, *in* which $n = 0.3$ is the porosity of the armour layer, ρ_C and ρ_W is the mass density of concrete and water, respectively, and $\varphi = 60^\circ$ is the estimated friction angle of the blocks.

The horizontal wave induced force transferred through the blocks to the wave wall is on the safe side set to $F_{HB} = 77$ *kN/m* corresponding to recorded typical maximum loads in the model tests.

5.3. *Failure mode limit state functions and statistical parameters*

5.3.1. *Sliding of superstructure*

The limit state function is written

$$g = (F_G - F_U) f - F_H,$$

in which F_G is the weight of the superstructure modelled by a Normal distributed multiplicative stochastic variable with mean value 1and COV 0.02, F_U is calculated from p_U – values given in Table 6, the friction factor with mean value $f = 0.7$ is assumed Log Normal distributed with COV = 0.15, and F_H is total horizontal load on the superstructure F_{HW} (see Table 6) plus F_{HD} and F_{HB}.

5.3.2. *Slip failure in core material*

Figure 17 shows the slip failure geometry. The effective width B_Z of the wave wall is determined such that the resultant vertical force $F_G - F_U$ is placed $B_Z/2$ from the heel of the base plate. The limit state function is written

$$g = Z(\gamma_R - \gamma_W) A\, \omega_V + (F_G - F_U)\, \omega_V - F_H\, \omega_H, \text{ in which}$$

Z is the model uncertainty assumed Log Normal distributed with mean value 1 and COV = 0.1, γ_R and γ_W are specific weights of core material and water, respectively, A is the area of the rupture zone, $\omega_V = \sin(\varphi_d - \theta) / \cos(\varphi_d)$ is the vertical displacement of the rupture zone, and $\omega_h = \cos(\varphi_d - \theta) / \cos(\varphi_d)$ is the horizontal displacement of the rupture zone. The effective friction angle of the core material φ_d is assumed Log Normal distributed with mean value 40° and COV = 0.075.

5.3.3. *Hydraulic stability of main armour, toe berm and rear slope*

The limit state functions are in the form $g = D_C - D(H_S)$ in which D_C is the critical damage levels given in Table 7. The damage level $D(H_S)$ is given by a piecewise linear functions for specific H_S obtained by linear interpolation using Tables 8, 9 and 10, which are obtained from the model tests.

Table 7. Critical damage levels for armour in terms of relative number of displaced units

Armour type	SLS	ULS
Main armour, $D_{C,A}$	5%	15%
Toe berm, $D_{C,T}$	5% - 10%	30%
Rear slope, $D_{C,B}$	-	2%

The reason for the very restrictive damage level for rear slope armour is the brittle failure development observed in the model tests.

Table 8. Mean values and standard deviations, σ, of armour damage levels in % for different significant wave heights

Armour	Main armour		Toe berm		Rear slope	
H_S (m)	Mean	σ	Mean	σ	Mean	σ
9			0.03	0.03		
10	0.025	0.029	0.50	0.50		
11	0.19	0.047	0.60	0.60		
12	0.43	0.087	1.13	0.41		
13	1.50	0.43	2.23	0.80		
14	2.83	0.77	3.33	1.04		
14.3					0	0
15	4.73	0.50	5.60	0.74		
15.2					1	0.2
16	7.73	0.50	8.3	1.0		
16.5					15	3

5.4. *Failure probability results*

Monte Carlo simulation technique was used to obtain the failure probabilities shown in Table 9 for the failure modes Sliding, Slip failure, Main armour damage initiation, Main armour failure, Toe berm damage initiation, and Toe berm failure. The values given in Table 9 relate to the superstructure cross sections shown in Fig. 17, having a wall thickness of 10 m and a cross sectional area of 185 m^2. H_S is modelled by a truncated Weibull distribution, cf. Fig. 3, and statistical uncertainty is included.

Table 9. Failure probabilities in 50 years reference period. Wave wall thickness 10 m. Superstructure cross sectional area 185 m^2

Sliding	Slip failure	Main armour damage SLS	Main armour failure ULS	Toe berm damage SLS	Toe berm failure ULS	Total ULS
0.0623	0.209	0.32	0.041	0.40	0.0011	0.21

The sensitivity of the failure probabilities for sliding and slip failure to the weight of the superstructure was investigated as well. If only the thickness of the wave wall is increased from 10 m to 12 m giving a cross sectional area of 209 m^2, then the failure probabilities for sliding and slip failure are reduced to 0.0356 and 0.174, respectively.

The failure probabilities for sliding and slip failure are conservative because the slip failure is modelled by a two-dimensional model, where a more realistic model should include three-dimensional effects. Moreover, a triangular uplift pressure distribution is assumed although the model tests show lower pressures.

On the background of the design target probabilities of damages given in ROM 0.2.90, cf. Section 3, it is seen that the design complies fairly close to the recommended safety level of P_f = 0.30 for damage initiation, but is much more safe than corresponding to the recommended safety level of P_f = 0.15 for total destruction.

Table 10 shows the failure probabilities if the rear slope failure mode is included. Three different levels of the critical damage $D_{C,B}$ were investigated.

Table 10. Failure probabilities in a 50 years reference period. Wave wall thickness 12 m. Cross sectional area of superstructure 209 m^2

$D_{C,B}$ %	Sliding	Slip failure ULS	Main armour damage SLS	Main armour failure ULS	Toe berm damage SLS	Toe berm failure ULS	Rear slope failure ULS	Total ULS
2	0.0356	0.174	0.32	0.041	0.40	0.0011	0.59	0.62
5	-	-	-	-	-	-	0.51	0.54
10	-	-	-	-	-	-	0.37	0.42

It is seen from Table 10 that the rear slope failure mode is very critical with failure probabilities in the order of 0.4–0.6. Because such high failure probabilities are not acceptable, the conclusion of the reliability analysis was that a new and stronger design of the rear slope had to be implemented in the otherwise acceptable cross section.

6. Final Design of the Trunk Section in 40 m Water Depth

The change from the initial anticipated type of crane, shown in Fig. 13, to freely moving crawler cranes, see Fig. 20, plus a demand for more space for pipeline installations led to a demand for a much wider platform behind the wave wall. A total width of app. 18 m was needed.

Moreover, in order to have the wave wall as a separate structure without stiff connection to a thick base plate, and in order to reduce the wall width from 10 m to a varying width of 5 to 10 m, the wave induced loadings on the wall were reduced by extending the width of crest of the main armour berm at level + 25 m by 4 m to 20 m. This solution was preferred despite the significantly larger volume of the structure.

Fig. 20. Liebherr LR 11350 crane

The final design of the cross section is shown in Fig. 21.

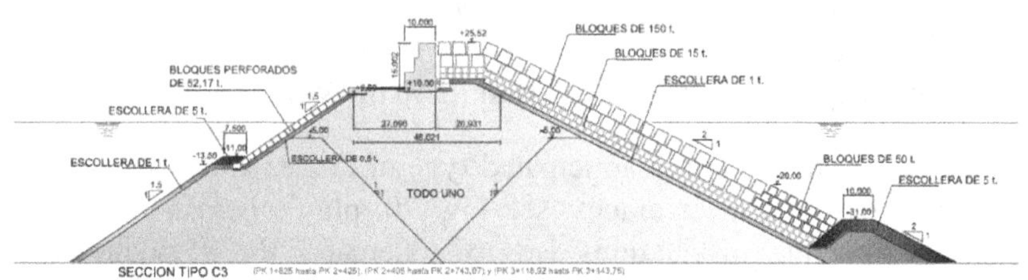

Fig. 21. Final cross section of trunk in 40 m water depth

The final trunk cross section design was developed mainly through large scale 2-D model test at Cepyc- CEDEX, Madrid. The length scales were 1:25 and 1:28.5.

A photo of model testing of the finally designed cross section is shown in Fig. 22. The 150 t cubes were placed randomly by a crane in two layers except for the regularly placed four rows of cubes in the top berm. From level –20 m down to the top of the toe berm in level –28 m the two layers of 150 t cubes were substituted by three layers of 50 t cubes (six rows) in order to comply with the capacity of the cranes. The relatively deep placement of the toe made it sufficient to use three layers of 5 t rock as toe armour.

Fig. 22. Photo of armour before testing in the scale 1:28.5 model tests at Cepyc-CEDEX

The sufficient stability of the main armour 150 t cubes was confirmed although some settlement of the armour took place in wave with H_S larger than 7 m. Figure 23 shows the upper part of the armour layer after exposure to design waves.

Fig. 23. Settlements in main armour layer of 150 t cubes after exposure to design waves with H_s = 15 m. From 1:28.5 scale tests at Cepyc-CEDEX, Madrid

The rear slope and the shoulder of the roadway behind the wave wall are very exposed to splash-down form the overtopping waves. The loadings from the splash-down were recorded by pressure transducers, and the stability of the rear slope extensively studied. The final solution was to place, as a pavement on slope 1:1.5 in a single layer, parallelepiped concrete blocks of 53 t with holes for reduction of the effect of the wave induced pore pressures in the core, cf. Fig. 35. The dimensions of the blocks are in meters: 5.00 x 2.50 x 2.00 with two holes of 1.00 x 0.70 x 2.00 each. The hollowed blocks are supported by a 7.5 m wide berm of 1 t rocks at levels ranging from –11.8 m to -13.5 m, and covered with one layer of 5 t rocks. The geotechnical stability of the rear slope was studied by slip circle analyses; cf. Fig. 24 which show an example of the loadings form the splash-down and the related slip failure safety factor denoted C.S.

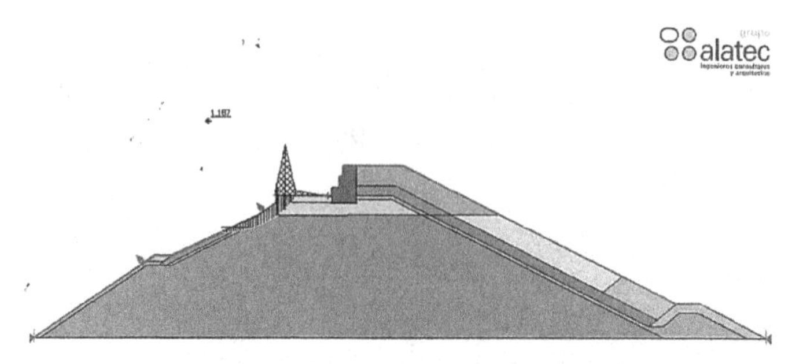

Fig. 24. Example splash-down loadings from overtopping waves and related slip failure safety factors for rear slope stability. From report of ALATEC

The large overtopping makes it necessary to protect the liquid bulk pipe gallery with a reinforced concrete roof attached to the wave wall as shown in Fig. 25. Moreover, an overhanging slab is added for the protection of the top of the rear slope.

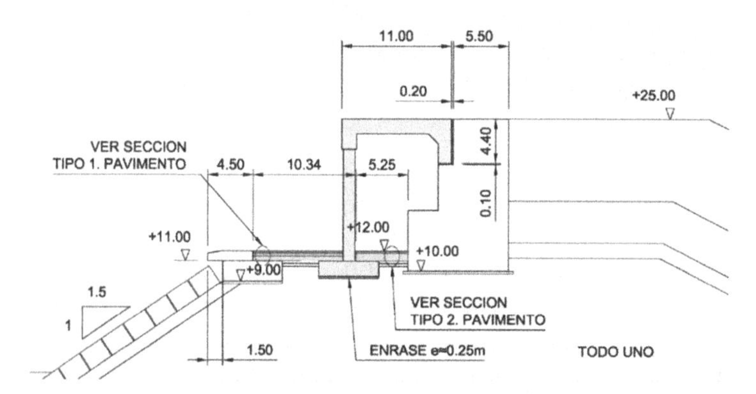

Fig. 25. Structure for the protection of pipes against slamming from overtopping waves.

The consultant ALATEC did a Level I deterministic analysis in accordance with ROM 0.5.05, [16] of the safety of the wave wall in the final cross section shown in Fig. 21 using the wave loadings recorded in the large scale model tests at Cepyc-CEDEX. The results are given in Table 11. The minimum safety coefficient as requested in ROM 0.5.05 is 1.1.

A Level III reliability analysis based on readily available formulae for structure responses is presented in Section 10.4.

Table 11. Wave wall safety coefficients. Level 1 analysis by ALATEC

Failure mode	Safety coefficient
Sliding	1.67
Tilting	2.34
Bearing capacity, method of Brinch Hansen, [15]	1.22
Global slip circle stability analyses	1.29

7. Design of the Roundhead and Adjacent Trunk

As seen from Figs. 2 and 29, a spur breakwater, 390 m long and made of caissons, separates the port basin from the outer part of the breakwater. The latter consists of a 240 m long trunk section without superstructure, and a roundhead. 3-D wave agitation model tests at The Hydraulics and Coastal Engineering Laboratory at Aalborg University, Denmark, were used to determine the optimum combination of trunk length and crest level considering target acceptable wave conditions in the area behind the breakwater.

The cross section of the trunk is shown in Fig. 26. Two layers of randomly placed 150 t cubes on slope 1:2 formed the sea side armour layer. A crest level of +17.35 m and a single layer of 150 t regularly placed cubes on the crest were selected as the most economical solution.

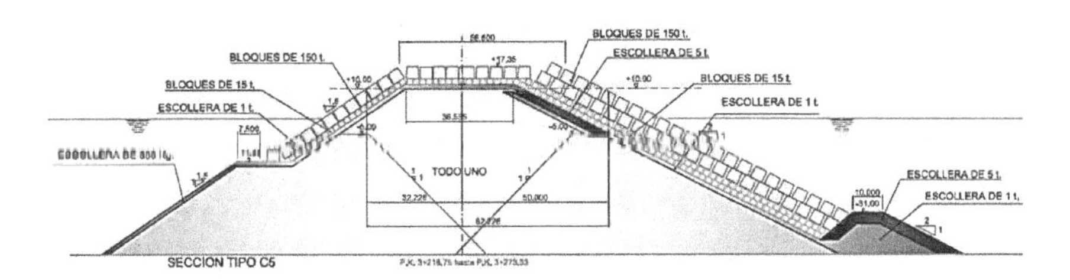

Fig. 26. Trunk cross section of the outer part of the main breakwater

The weight of randomly placed armour units in roundheads usually has to be approximately twice the weight of the units in the adjacent trunk. This however would be impossible because of the capacity limitations of the cranes. A solution was found by increasing the mass density of the concrete in the most critical sectors of the roundhead. The dimensions of 4 x 4 x 4 m for the 150 t cubes were maintained, but the mass density was increased to 2.80 t/m^3 and 3.04 t/m^3 resulting in masses of 178 t and 195 t respectively. The stability of the roundhead solutions using high mass density cubes was intensively tested in parametric 3-D model tests at the Hydraulics and Coastal Engineering Laboratory in Aalborg. Figure 27 shows photos from a test series in which white colored cubes of 4 x 4 x 4 m (prototype), having mass density 2.80 t/m^3, were placed only in the top layer in the most critical sector of the head. All other cubes had mass density 2.40 t/m^3.

Fig. 27. Model test of roundhead armored with 4 x 4 x 4 m (prototype) concrete cubes. Mass density of the white colored cubes is 2.80 t/m^3. Other cubes have mass density 2.40 t/m^3. After exposure to long test series with H_S up to 14.3 m. From model tests at the Hydraulics and Coastal Engineering Laboratory at Aalborg University, Denmark.

The efficiency of using high density unit is seen from Fig. 28.

Based on the parametric model tests a new formula for stability of cube armoured roundheads was developed (Macineira and Burcharth, 2004), [11].

$$\frac{H_s}{\Delta \cdot D_n} = 0.57 \cdot e^{0.07 \cdot R/Dn} \cdot \cot\alpha^{0.71} \cdot D_{\%}^{0.2} \cdot S_{op}^{0.4} + 2.08 \cdot S_{op}^{0.14} - 0.17$$

R is the radius of the roundhead at SWL, α = slope, D_n = side length of cubes, $D_{\%}$ = relative number of displaced cubes in percentage, S_{op} = deep water wave steepness, $\Delta = (\rho_s / \rho_W - 1)$ in which ρ_s and ρ_W are mass density of cubes and water respectively.

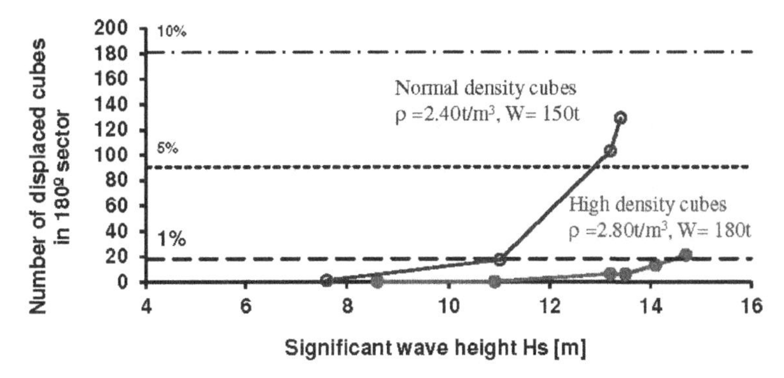

Fig. 28. Comparison of stability of normal and high mass density cubes in roundhead armour layer. From model tests at the Hydraulics and Coastal Engineering Laboratory at Aalborg University, Denmark

The finally chosen roundhead design is shown in Figs. 29 and 30.

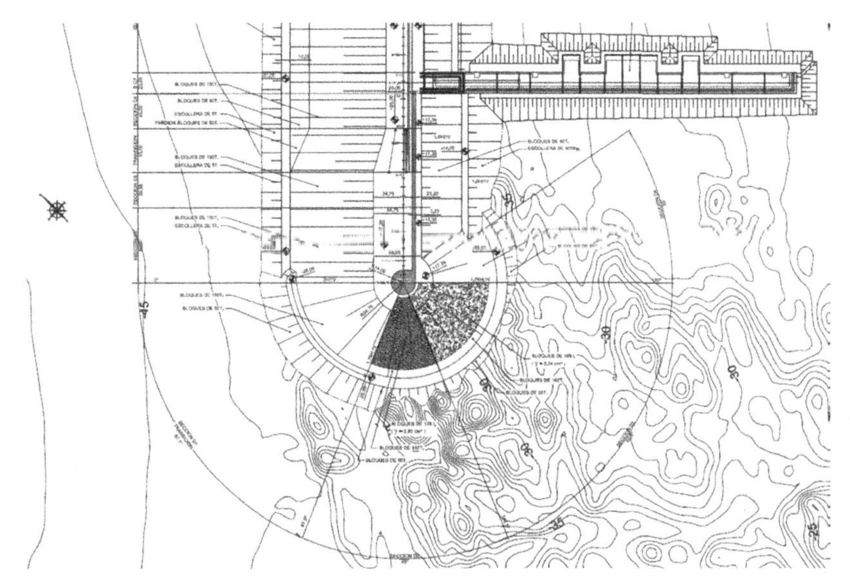

Fig. 29. Layout of the roundhead and spur breakwater

The slope of the roundhead is 1:1.75. Cubes of concrete mass density 2.80 t/m^3 and mass 178 t were used in the top layer in the dark shaded sector, and cubes with mass density 3.04 t/m^3 and mass 195 t were used in the top layer in the grey shaded sector down to level – 19.5 m. The two lower rows of cubes and the under layer consist of the 150 t cubes also used in the breakwater trunk. The filter layers are 15 t cubes and 1 t rocks except in the upper part where 5 t rocks are used. The toe berm armour is 50 t cubes.

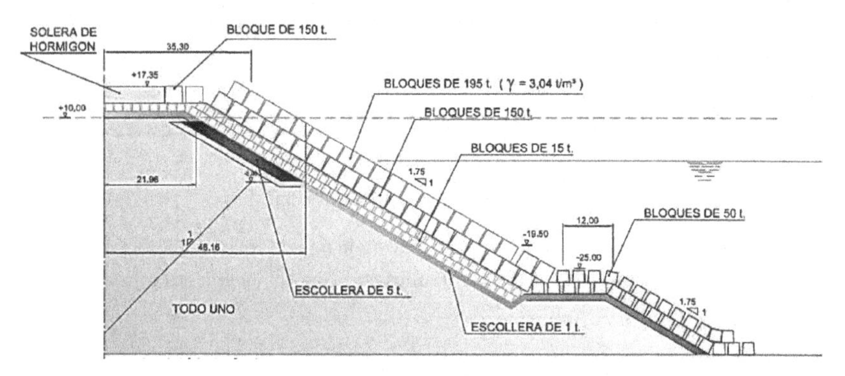

Fig. 30. Cross section of the roundhead

The final design of the roundhead was studied in verification model tests at Cepyc- CEDEX, and satisfactory performance was observed.

During the design phase a solutions for a breakwater head consisting of caissons were studied in 3-D model tests in short crested waves at the Aalborg University laboratory and at Cepyc-CEDEX. The large dimension of the rubble mound trunk and the large water depth made it necessary to use six adjacently placed caissons in the head. The very large wave forces and the related deformation-responses of the caissons made the consequences of the caisson interactions questionable. Moreover, difficulties in obtaining stability of the 150 t cubes in the transition between the caisson and the armour layer were observed. Also, the very strong vortices around the inner corner of the caissons made it very difficult to obtain sufficient stability of the foot-protection berm although very heavy concrete blocks were used (150 t normal density and at Aalborg University also 180 t high density concrete cubes as well as 150 t normal density and 180 t high density Japanese hollowed foot protection blocks). Besides this, if a caisson solution was chosen then the trunk had to be longer in order to obtain the same wave conditions behind the breakwater as in the roundhead solution.

8. Construction of the Breakwater

8.1. *Construction method and equipment*

The consortium UTE Langosteira, consisting of the companies DRAGADOS; SATO, COPASA and FPS, was awarded the contract for the works.

The construction of the breakwater started in the spring of 2007 and was completed in the autumn of 2011.

The wave climate at Punta Langosteira is very seasonal with frequent storms in the winter period from November to April. Figure 31 shows some wave statistics illustrating typical variations over the year. The sequences in the execution of the works reflected both the severe wave conditions and the seasonal variation. When weather conditions allowed barges to operate, dumping of rock material was possible throughout the year up to level –20 m over which level storms would flatten the dumped material. The procedure was to dump from split barges two moats of unhewn core material in strips along the two edges of the fill area. The outer slope of the moats was covered with a filter layer of 1t rocks, and on top of this two layers of 15 t cubes were placed on the seaside slope as part of the final structure. Core material was then dumped in between the armoured moats. This procedure was repeated in layers up to level –10 m. The top of the core material was unprotected below level –20 m during the winter seasons.

Dumping from floating material could proceed up to level –5 m but only during the summer campaign. From level –10 m to level –5 m it was not considered necessary to place edge moats as the risk of erosion was very limited due to the fast proceeding placement of protection layers placed by land based equipment, see Fig. 32. From level –5 m to the top of the fill only land based equipment was used.

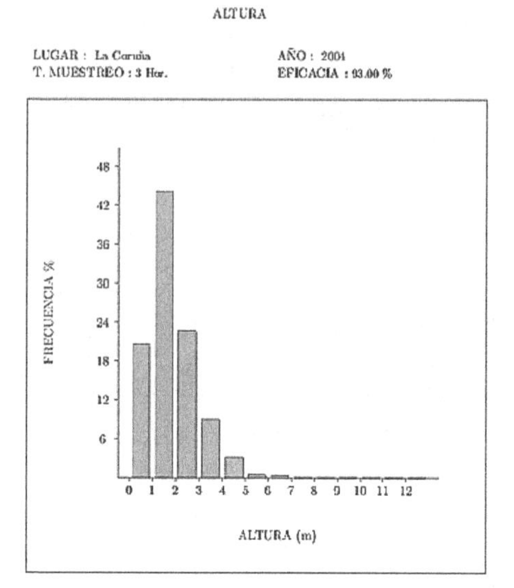

Alturas Máximas de Ola registradas en la boya de Langosteira (Septiembre 2005 - Agosto 2007)

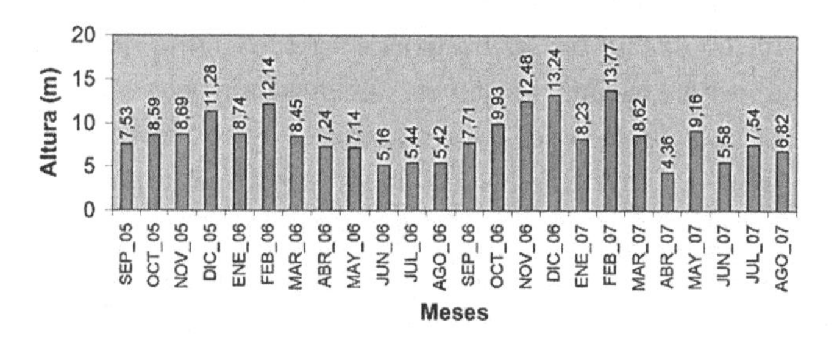

Fig. 31. Typical variations in wave heights at Punta Langosteira

The level of the working platform/cause way was as high as +10 m in order to avoid frequent dangerous overtopping. The unhewn quarry rock core material was placed by direct dumping from 90 t dumpers assisted by bulldozers. The filter layers of 1 t rocks were placed by crane using a tray with 40 t capacity per operation. The sea side under-layer consisting of 15 t cubes were placed by cranes following closely behind the placement of the filter layer. Finally, very closely behind, the large cranes placed the sea side main armour cubes of 70 t, 90 t and 150 t, depending on the section.

Fig. 32. Construction procedure with land based equipment

The two cranes used for placement of the 150 t cubes in the trunk and the 180 t and 195 t cubes in the roundhead were type Liebherr LR 11350, see Fig. 20. The capacity of the cranes was dependent on the boom/derrick and counterweight system. With a 135 m boom and 600 t counter weight the capacity corresponded to:

Placement in the dry:	Reach/radius 88 m,	Max. load 170 t
Under water:	Reach/radius 115 m,	Max. load 110 t

The cranes were able to place 30 cubes in 24 hours. Pressure clamps were used to hold the cubes. A GPS system using UTM coordinates was used for the positioning of the units. The units were released from the crane when touching the armour layer.

The under-water works were controlled and checked by a multi-beam echo sounder system operated from a boat; see the example plot in Fig. 33.

Fig. 33. Multi-beam echo sounder view of the breakwater under construction in the summer of 2008.

The cubes were transported on dumpers and trailers from the production yard to the cranes. Figure 34 shows the transport of three 150 t cubes.

Fig. 34. Transport of 150 t cubes

The main armour was placed before casting of the wave wall, cf. Figs. 35 and 36.

Fig. 35. Casting of the wave wall in lee of the placed 150 t main armour cubes resting on the under-layer of 15 t cubes.

Fig. 36. Casting of the wave wall

Figure 37 shows a photo of the completed wave wall. The two persons standing at the wall indicate the size of the structure.

Fig. 37. Photo of the completed wave wall.

Placement of the 53 t hollowed blocks under water on the rear slope of the breakwater was difficult as accurate positioning in a one layer pavement was needed. A special on-line GPS/SONAR system was developed and surely needed for this under-water operation. Figure 38 shows part of the completed rear slope armour.

Fig. 38. Rear slope consisting of 53 t hollowed parallelepiped concrete blocks placed in one layer as a pavement

Securing the vulnerable open end of the structure during the winter storms demanded construction of an interim roundhead armoured with 150 t cubes each autumn. Figure 39 shows the third interim roundhead for the winter 2009.

Fig. 39. Interim roundhead for the winter season 2009

The construction of the 2008 interim winter roundhead is seen in Fig. 40.

Fig. 40. Construction of interim winter roundhead armoured with 150 t cubes.

Despite great efforts in securing the outer end of the structure, severe damage occurred to the head during a storm in January 2008 with significant wave heights up to approximately 10.5 m, cf. Fig. 41.

Fig. 41. Damage to the winter roundhead in the very stormy January of 2008 with significant wave heights up to approximately 10.5 m.

8.2. *Production and storage of concrete cubes*

A high capacity plant for production of concrete and armour blocks was installed in 2007, see Fig. 42. The plant had two production lines. One for production of 15 t cubes with a capacity of 220 units per day. The other for the production of large cubes ranging from 50 t up to 195 t. The capacity corresponded to 32 cubes of 150 t per day. Each of the two production lines had a capacity of 2000 m^3 concrete per day.

The aggregates for the concrete mix were supplied from two mobile rock crushing plants each having a capacity of 150 t per hour. The rock material was supplied form the quarries in the rocks surrounding the port area.

Fig. 42. Overview of plant for production of concrete and armour blocks. Also seen is the interim construction port with a 350 m long block type quay protected by a 500 m long breakwater armoured with 5.880 cubes of 50 t.

The steel formwork was removed six hours after completion of pouring of the concrete, and eight hours later the cubes were lifted and moved by gantry cranes to the adjacent storage areas and stacked. A close up of the steel shutter in the casting yard is shown in Fig. 43.

Fig. 43. Steel shutters for the 4 x 4 x 4 m cubes in the casting yard.

The specifications for the concretes are given in Table 12.

Table 12. Concrete specifications

Constituents	Av. density (t/m^3)	Component's weight (kg per m^3 concrete)		
		150 t cubes	178 t cubes	195 t cubes
Cement II/B-V 32.5R MR, Portland, low heat, sulfate res., siliceous fly ash content	3.20	340	310	310
Granite and quarzitic sand. Max. size 70 mm	2.59-2.65	1898		
High density aggregates. Max. size 40 mm. Amphibolite Barite	2.96-3.06 4.02-4.17		2273	1000 1648
Viscocrete 3425 or Melcret 500		2	2	2
Water		160	140	160
Water- Cement ratio		0.47	0.45 - 0.5	0.515
f_c ,90 days/182 days (Mpa), average strenght obtained		39	34.4/ 39.7	33.2/ 40.8
Cube (t/m^3), average mass density obtained		2.37	2.70	3.05
Aggregate gradation		According to the Fuller formula		

A number of 4 x 4 x 4 m high density concrete cubes suffering from ultrafine cracks in some surface areas zones was produced during the trial casting. The problem was most probably due to thermal cracking caused by high temperature differences (exceeding 20°C) during the hardening process.

The number of cubes suffering from thermal cracks was as follows:

108 of mass density 2.78 t/m3
11 of mass density 3.04 t/m3

It is supposed that these cubes have a somewhat reduced strength for which reason they had to be placed in areas in which no movements (rocking or displacements) were expected. The one-layer horizontal crests of the head and of the adjacent low-crested trunk in which the cubes are placed as a pavement with crest level + 17.4 m were selected; see Figs. 25 and 29.

These optimal areas for the placement were identified from the series of physical model tests performed at the Hydraulics and Coastal Engineering Laboratory, Department of Civil Engineering, Aalborg University, Denmark.

8.3. *Volume and costs of the works*

The overall figures for the volumes used in the main breakwater are as follows:

Unhewn rock materials, mainly for the core: 14.538 million m^3.
Sorted rock materials, mainly for filter layers: 8.219 million m^3.
Concrete in wave wall and road pad: 0.368 million m^3.
Concrete in placed cubes: 2.891 million m^3.

Table 13. Number of concrete cubes placed in the main breakwater

Mass (t) of cube	15	50 -53	70	90	150	178	195	Total
Number	116.307	14.835	9.145	1.874	22.755	197	276	165.389

The total cost of the breakwater including the spur breakwater is approximately 384.360.000 EUR including 17% VAT.

9. Quality Assurance of the Works

In order to control the quality and safety of the works, the organization shown in Fig. 44 was established.

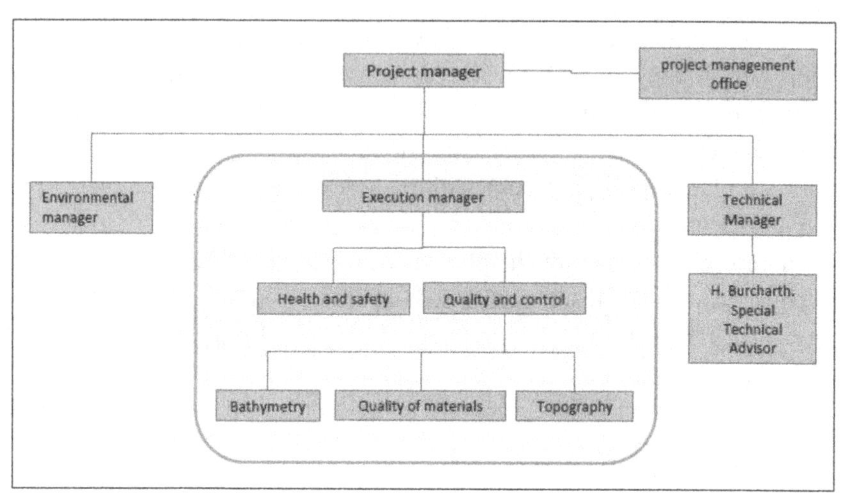

Fig. 44. Organization of the quality control and safety of the works

The more important elements in the quality assurance were as follows:

The control of the rock materials properties in terms of gradation and friction angle of the production from quarries and crushing plants was done on a daily basis.

The quality of concrete in terms of strength, mass density and temperature was checked from samples of all batches.

All concrete blocks were assigned x-y-z coordinates and placed accordingly by the use of GPS equipment mounted on the crane. Every vessel placing material was positioned by GPS and all movements were registered. Final locations of under-water armour and slope surface geometry were checked by multi beam echo-sounder.

Every day the environmental conditions in terms of wave height and wave period, wind velocity, rain and temperature were forecasted for the following three and seven days, and related wave run-up and overtopping were calculated. The works were planned accordingly considering the risk of human injury and damage to equipment.

A plan for checking of archaeological effects was implemented.

10. Reliability Analysis of the Completed Breakwater

10.1. *Introduction*

A reliability analysis of the completed cross section shown in Fig. 21 and of the completed roundhead shown in Figs. 29 and 30 was performed, solely based on available formulae for wave loadings and armour stability. Thus, the basis for the analysis differs from the reliability analysis of the initial design which was based on model tests results, cf. Paragraph 5.

10.2. *Stochastic models*

10.2.1. *Significant wave heights and wave periods*

A three parameter Weibull distribution was fitted by Maximum Likelihood method to H_s-values exceeding 7.20 m recorded by the Langosteira buoy in the years 1998 – 2011, a period of 13.8 years:

$$F(H_s) = (1 - exp(-((H_s - \gamma)/\beta))^{\alpha})^{\lambda}, \qquad H_S = \gamma + \beta\left(-\ln\left(\frac{1 - F(H_S)}{\lambda}\right)\right)^{1/\alpha}$$

with α = *0.96*, β =*1.09*, γ = *7.15* and λ = *2.54*. The fitting and the related return period H_s-values are shown in Fig. 45.

The uncertainty of the recorded H_s-values is assumed corresponding to a standard deviation of $\sigma(H_s) = 0.05$. The statistical uncertainties related to $F(H_s)$ of the return period significant wave heights, $\sigma(H_s^T)$, were estimated by the method of Goda, see [12]. The resulting standard deviation is taken as $(\sigma(H_s)^2 + \sigma(H_s^T)^2)^{0.5}$.

Based on analysis of the buoy data the following relation was found for the spectral peak period $T_P(H_s) = 9.26\, H_s^{0.22}$, Normal distributed with COV = 6%. The average of the mean wave period was determined to $T_z = T_p$ /1. 20 with a range of $T_z = T_p/1.05 - T_p/1.35$. For the actual calculations a ratio of $T_z = T_p/1.10$, corresponding to typical ocean wave conditions was used.

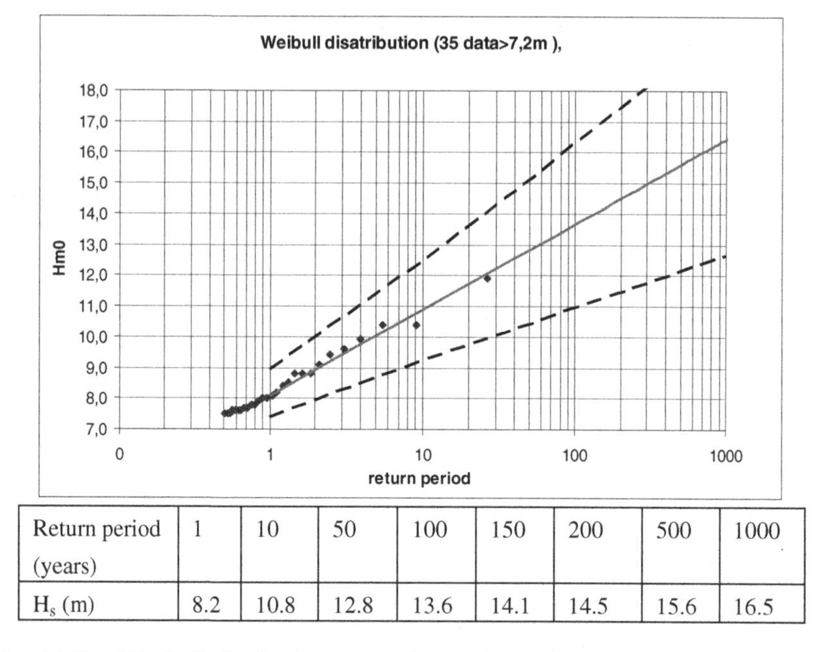

Return period (years)	1	10	50	100	150	200	500	1000
H_s (m)	8.2	10.8	12.8	13.6	14.1	14.5	15.6	16.5

Fig. 45. Fitted Weibull distribution with 90% confidence bands and related return period H_s-values

10.2.2. *Model for tidal elevations*

See Paragraph 5.2.3.

10.2.3. *Modelling of wave forces on wave wall*

The wave forces were estimated by the method of Pedersen, [10] and [14], modified by Noergaard et al., [17], to cover also intermediate and shallow water waves. The coefficients in the modified Pedersen formulae are assumed Normal distributed with COV-values as given in [17]. The 0.1% wave height used in the Noergaard equation was determined by the Battjes-Groenendijk method, [18] with input of deep water spectral significant wave height, water depth and sea bed slope.

10.2.4. *Material and geometrical parameters*

Cube side lengths: Normal distributed with COV = 1%.

Armour slopes: Mean slope for trunk, 1:2, Normal distributed with COV = 5%. Mean slope for roundhead, 1:1.75, Normal distributed with COV = 5%.

Concrete mass density: Normal distributed with mean value 2.35 t/m^3 for the trunk, and 3.04 t/m^3 for the roundhead. In both cases with COV = 1%.

Friction coefficient in interface between core material and wave wall: mean value of 0.7, Normal distributed with COV 0.15.

Friction angle of core material: Mean value of 37.5°, Normal distributed with COV = 11%.

10.3. *Failure mode limit state functions*

10.3.1. *Stability of cube armour layers in trunk, roundhead and toe berms*

The stability of the trunk cube armour is estimated by the formula of Van der Meer, [9], valid for slope 1:1.5 but modified to cover slope 1:2:

$$N_s = H_s/(\Delta D_n) = (2/1.5)^{1/3}\,(6.7 N_{od}^{\,0.4}/N_z^{\,0.3} + 1.0)\, s_m^{\,-0.1}$$

N_{od} is the relative number of displaced units with in a strip of width D_n. N_z is the number of waves, and s_m is the deep water wave steepness using the mean wave period. The model uncertainty of the formula is given by a Normal distributed factor of mean value 1.0 and COV = 0.1.

For the cube armoured toe berms the formula of Burcharth et al. (1995), [13] and [14] is used although developed for parallelepiped blocks:

$$N_s = H_s/(\Delta D_n) = (0.4\, h_b/(\Delta D_n) + 1.6)\, N_{od}^{\,0.15}$$

h_b is the water depth over the toe. The model uncertainty of the formula is given by a Normal distributed factor of mean value 1.0 and COV = 0.1.

For the stability of the cube armour in the roundhead the formula of Maciñeira and Burcharth, [11], is used, see Paragraph 7. The model uncertainty of the formula is given by a Normal distributed factor of mean value 1.0 and COV = 0.06 applied to the right hand side of the equation.

The limit state functions are in the form, g = $N_{od,C} - N_{od}$ (H_s) in which $N_{od,C}$ is the critical damage levels corresponding to the design limit states as given in Table 14.

Table 14. Critical damage levels in terms of $N_{od,C}$ for cube armour in trunk, roundhead and toe

Armour type	SLS	ULS
Trunk armour	$N_{od,C}$ = 1.3 (D = 5%)	$N_{od,C}$ = 4.0 (D = 15%)
Toe berm in trunk	$N_{od,C}$ = 2.0 (D = 5 - 10%)	$N_{od,C}$ =4.0 (D = 30%)
Roundhead armour (most critical 45^0 sector)	D= 2%	D = 10%

The ULS geotechnical stability of the wave wall is investigated for two failure modes, a plane slip failure as shown in Fig. 18 and a local bearing capacity slip failure. The effective unit weight of the core material is set to 10.7 kN/m^3, and the friction angle to 37,5° with COV 11%. The formula of Brinch Hansen, [15], (also given in [14]), with COV 10% is used for estimation of the soil bearing capacity.

10.4. *Results of the reliability evaluation*

The occurrence of the storms was assumed following a Poisson distribution. From Monte Carlo simulations using 10,000 randomly chosen H_s –values the failure probabilities within 50 year service life given in Tables 15 and 16 were obtained.

From Tables 15 and 16 is seen that the structure fulfill the requirements of ROM 0.2.90, [5], see Chapter 3. It should be noted that the failure probability of 0.15 given in [5] is related to total destruction, whereas the damage used for ULS in Tables 15 and 16 is related to severe damage.

Table 15. Trunk failure probabilities in 50 year service life

Structure element	Armour	Toe berm	Wave wall				Total
Failure mode	Displacement	Displacement	Sliding	Tilt	Bearing capacity	Slip circle	
Equation Reference	Van der Meer [9]	Burcharth [13]	Pedersen [10]	Pedersen [10]	Brinch Hansen [15]	Bishop	
Failure prob. SLS Return period (years)	0.34 120	0.18 247					0.36 113
Failure prob. ULS Return period (years)	0.066 738	0.069 699	0.01 4242	0.01 8434	0.02 2208	0.05 897	0.13 373

Table 16. Roundhead failure probabilities in 50 year service life

Structure element	Armour	Toe berm	Total
Failure mode	Displacement	Displacement	
Equation, Reference	Macineira [11]	Burcharth [13]	
Failure prob. SLS Return period (years)	0.15 298	0.17 267	0.20 221
Failure prob. ULS Return period (years)	0.04 1207	0.06 748	0.07 681

11. Performance of the Breakwater

Since completion in the autumn of 2011 and until end of March the breakwater has experienced the following storms given in Table 17.

Table 17. Storms since completions of the breakwater

Year	Month	Date	H_s (m)	T_p (s)
2011	2	15	9.9	16.6
-	12	12	8.1	16.6
2012	4	18	7.2	14.3
2013	1	28	7.5	16.6
-	10	28	6.5	16.9
-	12	25	7.7	14
2014	1	6	7.8	19
-	1	28	7.6	16
-	2	2	8.0	19
-	2	5	7.5	18
-	2	8	8.0	18
-	3	3	10.4	19.9

Fig. 46. Cavity in upper shoulder of armour layer due to compaction of front slope armour.

So far no damage has been observed to any part of the structure. The positions of the armour blocks are kept under observation by airborne photographical technique on a regular basis. In the outer low- crested part of the breakwater, see Fig. 26, some settlements of cubes due to compaction of the front slope armour occur leaving openings in the transition between the single layer and the double layer of cubes, see Fig. 46. The cavities will be filled with concrete.

References

1. Carci, E., Rivero, F. J., Burcharth, H.F., Macineira, E. (2002).The use of numerical modelling in the planning of physical model tests in a multidirectional wave basin. Proc. 28th International Conference on Coastal Engineering, Cardiff, Wales, 2002. pp 485-494, Vol. 1. Editor Jane McKee Smith.
2. Sorensen, J.D, Burcharth, H.F. and Zhou Liu (2000). Breakwater for the new port of La Coruna at Punta Langosteira. Probabilistic Design of Breakwater. November 2000. Report of Hydraulics and Coastal Engineering Laboratory, Aalborg University, Denmark.
3. AsistencianTecnico para la Redaccion de los Proyectos de Ampliacion de las Nuevas Instalaciones Portuarias de Punta Langosteira. Estudios Generales. SENER report P210C21-00-SRLC-AN-0003. Clima Maritimo.
4. Episidio de Meteorologia Maritima. UTE Langosteira Supervision. Report, November 9, 2010. Anexos 1, 2, 3 and 4.
5. ROM 0.2.90 (1990). Actions in the design of harbour and maritime structures. Puertos del Estado, Ministerio de Formento, Spain.
6. ROM 0.0 (2002). Recommendations for Maritime Structures. General procedure and requirements in the design of harbour and maritime structures. Part 1. Puertos del Estado, Ministerio de Fomento, Spain.
7. ROM 1.0.09 (2010). Recommendations for the design and construction of breakwaters, Part 1. Puertos del Estado. Minesterio de Formento, Spain.
8. Burcharth, H.F., Liu, Z. and Troch, P. (1999). Scaling of core material in rubble mound breakwater model tests. Proceedings of the 5th International Conference on Coastal and Port Engineering in Developing Countries (COPEDEC V), Cape Town, South Africa.
9. Van der Meer, J.W. (1988). Stability of cubes, Tetrapods and Accropode. Proceedings of Breakwaters'88, Eastbourne,U.K.
10. Pedersen, J. (1996). Wave forces and overtopping on crown walls of rubble mound breakwaters. Ph.D. Thesis, Hydraulic and Coastal Engineering Laboratory, Department of Civil Engineering, Aalborg University, Denmark.
11. Macineira, E. and Burcharth, H.F. (2007). New formula for stability of cube armoured roundheads. Proc. Coastal Structures, Venice, Italy.
12. Goda, Y. (2008). Random seas and design of maritime structures, 2nd edition, World Scientific, Singapore.
13. Burcharth, H.F., Frigaard, P., Uzcanga, J., Berenguer, J.M., Madrigal, B.G. and Villanueva, J. (1995). Design of the Ciervana Breakwater, Bilbao. Proc. Advances in Coastal Structures and breakwaters Conference, Institution of Civil Engineers, London, UK, pp 26-43.

14. Coastal Engineering Manual (CEM), (2002), Part VI, Chapter 5, Fundamentals of design, by Hans.F. Burcharth and Steven A. Hughes, Engineer Manual 1110-2-1100, U.S. Army Corps of Engineers, Washington D.C., U.S.
15. Brinch Hansen, J. (1970). A revised and extended formula for bearing capacity. Bulletin No 28, Danish Geotechnical Institute, Denmark.
16. ROM 0.5.05, (2005). Geotechnical recommendations for design of maritime structures. Puertos del Estado, Ministerio de Formento, Spain.
17. Noergaard, J. Q.H., Andersen, T.L. and Burcharth, H.F. (2013). Wave loads on rubble mound breakwater crown walls in deep and shallow water wave conditions, Coastal Engineering, 80, pp 137-147.
18. Battjes, J.A. and Groenendijk, H.W. (2000). Wave height distribution on shallow foreshores, Coastal Engineering, 40, pp 161-182.

CHAPTER 3

PERFORMANCE DESIGN FOR MARITIME STRUCTURES

Shigeo Takahashi

Port and Airport Research Institute
3-1-1,Nagase, Yokosuka, Japan
E-mail: takahashi_s@pari.go.jp

Ken-ichiro Shimosako

Port and Airport Research Institute
E-mail:Shimosako@pari.go.jp

Minoru Hanzawa

Fudo Tetra Cooporation
7-2, Nihonbashi-Koami-Chou, Chuou-ku,Tokyo, Japan
E-mail: minoru.hanzawa@fudotetra.co.jp

Reliability design considering probabilistic nature is quite suitable for maritime facilities because waves are of irregular nature and wave actions fluctuate. However, solely considering the probability of failure is considered insufficient, as deformation (damage level) should also be taken into account. This paper discusses performance design as a future design methodology for maritime structures, focusing on, for example, deformation-based reliability design of breakwaters and how it can be applied to stability design. The performance of breakwaters is specifically considered by describing the design criteria (allowable limits) with respect to different design levels and accumulated damage during a lifetime including probabilistic aspects. The new frame of performance design for maritime structures and necessary studies to develop the design are also discussed.

1. Introduction

Breakwaters are designed such that they will not damaged by design storm waves, although there is always a risk that waves higher than the design storm waves will occur. For example, in 1991, very high waves exceeding the design level led to failure of a vertical breakwater as shown in Photo 1, where the caissons slid until some fell off the rubble foundation. However more than half of the caissons were remain intact. Since failures are not considered in the current design process, coastal structure engineers do not pay sufficient attention to the extent and consequences of failure, i.e., performance evaluation is limited to the time prior to failure, while that during and after failure is neglected.

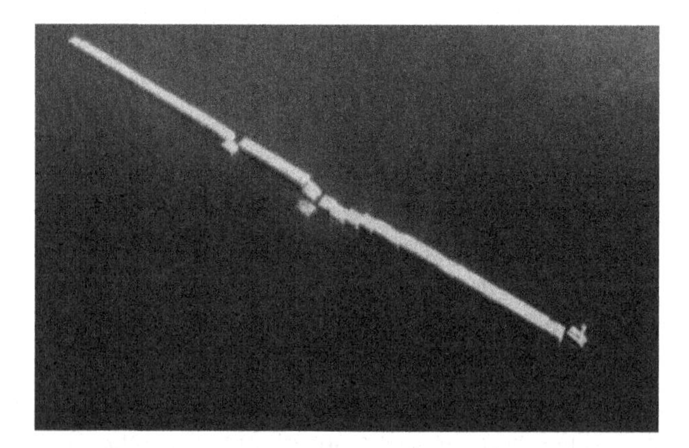

Photo 1 Caisson sliding caused by a typhoon.

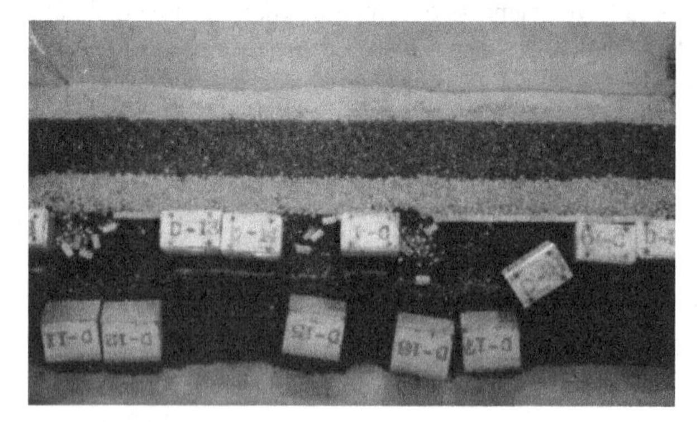

Photo 2 Model experiment showing caisson sliding.

Photo 2 shows the failure of vertical breakwater in a model experiment in which waves larger than the design one were applied to examine breakwater failure characteristics. Experiments like these are not required by Japanese design codes, yet they are frequently conducted for important structures such as offshore breakwaters since it is essential that designers know what actually occurs during the failure. In other words, not only the stability of the structure confirmed, but robustness/ resilience and cost aspects are examined as well.

Such experience has led us to conclude that the design of maritime structures needs a new design approach or "framework." And, because various civil engineering fields have recently focused on performance-based design, we feel this is an ideal concept from which to base the design of maritime structures in the 21st century. Here, we introduce deformation-based reliability design of breakwaters as an example of recently developed performance design, briefly examining this new frame of performance design for maritime structures.

The contents in this chapter are based on the papers published in the International Workshop on Advanced Design of Maritime Structures in the 21st Century (Takahashi et al., 2001) and in the Coastal Structures Conference on Coastal Engineering in Portland (Takahashi et al., 2003)

2. Performance Design Concept

2.1. *History and Definition of Performance Design*

While performance design started in Europe in 1960's, it became popular in the United States following the 1994 Northrige earthquake disaster. Stability performance of buildings and civil engineering structures is assessed in the performance design (SEAOC,1995).

The concept of performance design is new and fluid, which allows researchers and engineers to create an integrated design framework for its development. Performance design can be considered as a design process that systematically and clearly defines performance requirements and respective performance evaluation methods. In other words, performance design allows the performance of a structure to be explicitly and concretely described.

2.2. *New Framework for Performance Design*

Table 1 lists four aspects that should comprise the basic frame of performance design: selection of adequate performance evaluation items, consideration of importance of structure, consideration of probabilistic nature, and consideration of lifecycle. Regarding deformation-based reliability design, the four aspects are applied and will be explained below.

Table 1 Necessary considerations for performance design.

Selection of adequate performance evaluation items
Stability performance → Deformation (sliding, settlement, etc.) Wave control performance → Wave transmission coefficient, Transformation of wave spectrum Wave overtopping prevention performance → Wave overtopping rate, height of splash, inundation height
Consideration of importance of structure
Rank A,B,C → performance level
Consideration of probabilistic nature
Limit states design(Deterministic) design levels/limit states → performance matrix → toughness/repairable *Limit states Design with Level-3 reliability design* Risk/reliability analysis → probabilistic design/reliability design
Consideration of lifecycle
Lifetime performance→ accumulated damage/repair/ maintenance

2.2.1. *Selection of adequate performance evaluation items*

In the conventional design process, the stability of breakwater caissons is for example judged using safety factors of sliding and overturning, etc. based on balancing external and resisting forces. Deformation such as sliding distance, however, more directly indicates a caisson's stability performance.

Design based on structure deformation is not new for maritime structures, i.e., in conventional design the deformation of the armor layer of a rubble mound is used as the damage rate or damage level. For example, van der Meer (1987) proposed a method to evaluate necessary weight of armor stones and blocks which can also evaluate the damage level of the armor, while Melby and Kobayashi (1998) proposed a method to evaluate the damage progression of armor stones, and we proposed one to determine the damage rate of concrete blocks of horizontally composite breakwaters (Takahashi et al., 1998).

2.2.2. *Consideration of importance of structure*

Importance of the structure is considered even in current design. However, the performance design should deal with the structure's importance more systematically, and therefore the different levels of required performance should be defined in the performance design. The performance matrix can be formed considering the importance of the structure and probabilistic nature as will be described next.

2.2.3. *Consideration of probabilistic nature*

Probabilistic nature should be considered in the performance design of maritime structures because waves have an irregular nature and wave actions fluctuate. "Two" methods now exist to consider probabilistic nature.

a) Limit states design with a performance matrix

Firstly, a design should include performance evaluation against external forces which exceed the deterministic design value, e.g., stability performance against a wave with a 500-yr recurrence interval is valuable information leading to an advanced design considering robustness/ toughness and reparability. This probabilistic consideration was examined after the Northridge earthquake which is the foundation of anti-earthquake performance design. Current design codes for harbor structures in Japan include two design levels: a 75- and several hundreds-yr recurrence intervals. Here, the method considering probabilistic nature is named the "limit state design with a performance matrix."

b) Limit states design with Level-3 reliability design
Secondly, a performance design should include risk/reliability evaluation, which is called probabilistic or reliability design. With regard to reliability design, three levels exist: Level 1 uses partial safety coefficients as the limit state design for concrete structures; Level 2, the next higher level, uses a reliability index which expresses the safety level in consideration of all probabilities; while Level 3, the highest level of reliability design, uses probability distributions at all design steps.

Although the partial safety coefficient and reliability index methods (levels 1 and 2) are easily employed in standard designs, the probability distribution method (level 3) more directly indicates a structure's safety probability and is accordingly more suitable for a design method that considers deformation (damage level). Therefore, in performance design, Level 3 reliability design including deformation probability is necessary. Research has been performed in this area. Brucharth (1993), Oumeraci et al. (1999), and Vrijling et al. (1999) proposed partial coefficients methods for caisson breakwaters, van der Meer (1988) discussed a level 3 reliability design method for armor layers of rubble mound breakwaters considering the damage level, and Takayama et al. (1994) and Goda et al. (2000) discussed a level 3 reliability design method for caisson breakwaters that did not consider deformation (sliding).

2.2.4. *Consideration of lifecycle*

New designs should be extended to include life-cycle considerations, and a performance design should elucidate all performance aspects over structure lifetime. In conventional maritime design, only a short period is considered for exceptional waves attacking a breakwater although the construction period is considered when necessary. Since a structure performs during its design working time (lifetime) and therefore it accumulated damage during the lifetime should be considered.

In addition, its deterioration and maintenance should be considered in the design stage. While fatigue is included in the conventional design process, maintenance is not.

3. Deformation-Based Reliability Design for Caisson Breakwaters

3.1. *Performance Evaluation Method*

3.1.1. *Fundamental considerations of deformation-based reliability design*

A new performance design for the stability of breakwater caissons will be explained, being called deformation-based reliability design. Sliding distance is selected as the performance evaluation item and the probabilistic nature is fully considered. Performance design requires a reliable performance evaluation method. Thus, in deformation-based reliability design of a breakwater caisson, we developed a calculation method to determine the sliding distance due to wave actions, employing Monte Carlo simulation to include the probabilistic nature of waves and response of the breakwater caisson.

3.1.2. *Calculation method of sliding distance*

a) Deterministic value for a deepwater wave with a recurrence interval
Table 2 shows the flow of the calculation method of sliding distance due to a deepwater wave with particular recurrence interval (Shimosako and Takahashi, 1999). After specifying the deepwater wave, the incident wave is calculated by the wave transformation method (Goda 1973, 1985), providing not only the significant wave height but also the wave height distribution. For each wave, wave pressure distribution is evaluated and total horizontal and vertical wave forces are obtained with components, i.e., the standing wave pressure component α_1 and breaking/impulsive-breaking wave pressure component α_2/α_I (Takahashi, 1996). The time profiles of these components are sinusoidal and triangle, respectively.

The resisting forces against sliding are easily obtained from its dimensions, and the resisting force due to the movement of the caisson, being called the wave-making resisting force, can be formulated using the caisson dimensions. The equation of motion of the caisson with the external and resisting forces gives the motion of the caisson and resultant sliding distance. The equation considers only two-dimensional phenomena and is solved numerically.

Table 2 Flowchart for calculating sliding distance.

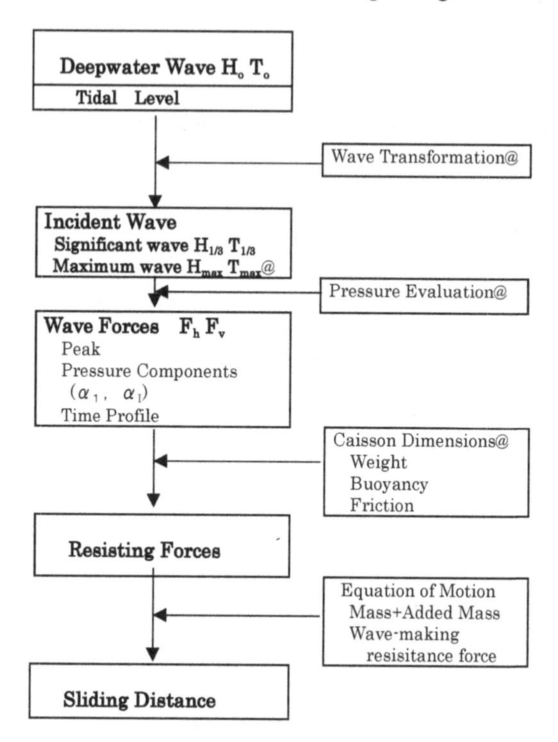

b) Probabilistic value for a deepwater wave of a recurrence interval

Even for a fixed deepwater wave condition, resultant sliding usually fluctuates due to the probabilistic nature of a group of waves and the response of the caisson. To obtain the probabilistic sliding distance for a given deepwater wave, fluctuation of the items denoted by @ should be considered. Table 3 shows parameters considered to reflect probabilistic nature in the present calculations and indicates bias of mean values and standard deviation (variance) of the probability distribution. The Monte Carlo simulation allows calculating the probability distribution of the sliding distance, with the calculation being repeated more than 5000 times from wave transformation to determination of the sliding distance for a fixed deepwater wave condition. From the probability distribution, the mean and 5% exceedance value are selected to represent the calculated distribution.

Table 3 Estimation errors for design parameters.

Design parameters	Bias of mean value	Variance
Wave Transformation	0.	0.1
Wave Forces	0.	0.1
Friction Coefficient	0.	0.1
Caisson Weight	0.	0.
Deepwater wave	0.	0.1
Tidal level	0.	0.1

c) Accumulated value during lifetime

To obtain the accumulated sliding distance during caisson lifetime (50 yr), one needs to consider the probabilistic nature of the deepwater wave and tidal level. The Weibull distribution with k = 2.0 is assumed as the extreme wave distribution with estimation error of 10% standard deviation. The tidal level is assumed as a triangle distribution between the L.W.L. and H.W.L. with error of 10% standard deviation. A total of 50 deepwater waves are sampled and the sliding distance is evaluated by the Monte Carlo simulation using the procedure in Table 2. Total sliding distance due to the 50 deepwater waves is the accumulated sliding distance for a 50-yr lifetime. The probability distribution of the accumulated sliding distance is obtained by repeating the calculations more than 5000 times. The accumulated sliding distance for a lifetime is called here 'accumulated sliding distance or lifetime sliding distance.'

3.2. *Example of Performance Evaluation*

3.2.1. *Design condition of a typical vertical breakwater*

Figure 1 shows a cross section of a composite breakwater designed against a design deepwater wave of a 50-yr recurrence interval of $H_{1/3}$ =

9.2 m and $T_{1/3}$ = 14 s with water depth h = 20 m (H.W.L. = +2.0 m). The caisson has width B = 23.68 m corresponding to a sliding safety factor SF = 1.07. The stability performance of the caisson, considered the sliding distance here, is explained next.

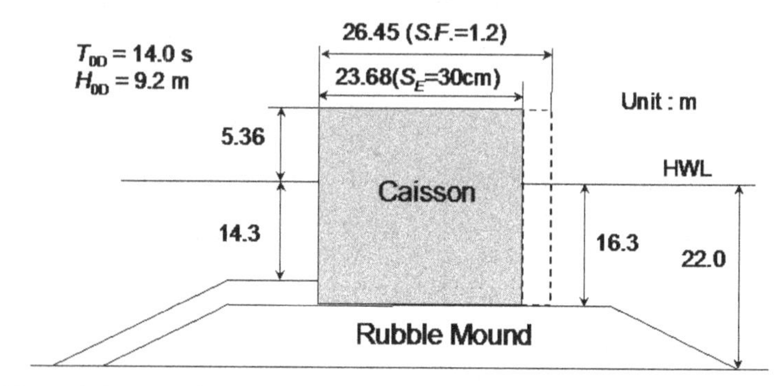

Figure 1 Cross section of a composite breakwater for performance evaluation example.

3.2.2. *Deepwater wave and sliding distance (deterministic value)*

Figure 2 shows caisson sliding distance produced by deepwater waves of different recurrence intervals, where the deterministic value of the sliding distance denoted by ■ is almost zero when the design wave with a 50-yr recurrence interval attacks. This result is considered quite reasonable since the safety factor for sliding is greater than 1.0. Note that even though deepwater wave height increases, sliding distance does not because the incident wave height is limited by wave breaking; i.e., the maximum incident wave height for a 50-yr recurrence interval deepwater wave is 15.07 m and only 10% higher at 16.56 m for a 5000-yr one.

3.2.3. *Deepwater wave and sliding distance (probabilistic value)*

Figure 2 also shows sliding distance due to deepwater waves obtained from the Monte Carlo simulation that included fluctuation of waves and sliding response of the caisson, where the mean (◆) and 5 % exceedance (↑) values of sliding distance are indicated. Due to the probabilistic nature, i.e., the occurrence of larger incident wave height and larger sliding, even the mean value of the sliding distance is greater than the deterministic values. In fact, the 5% exceedance value is much larger than the mean value. For example, for a wave with a 50-yr recurrence interval the mean value of the sliding distance is 7 cm and the 5% exceedance value is 17 cm, whereas for a wave with a 500-yr recurrence interval the values are 23 and 88 cm, respectively. Obviously then, the probabilistic nature must be considered.

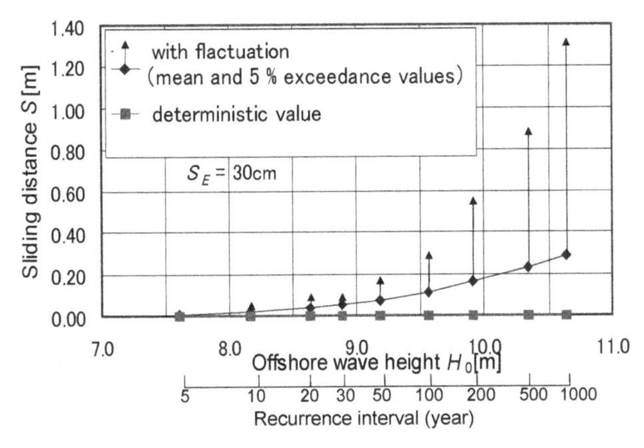

Figure 2 Deepwater wave height/recurrence interval vs. sliding distance (deterministic value).

3.2.4. *Probability of exceedance for a deepwater wave of N-yr recurrence interval over life time*

Figure 3 shows the probability of exceedance of the occurrence of the deepwater wave over a 50-yr life-time (design working time) vs. the recurrence interval of the deepwater wave. Since the estimation error in the Weibull distribution is considered to be 0.1 (variance), the probability of exceedance for the wave of a 50-yr recurrence interval is > 80%, being high compared to the conventional value of 63%. Even for the wave of a 500-yr recurrence interval the exceedance is still high, nearly 30%. For the wave of 5000-yr recurrence interval the probability is still nearly 10%. Considering the occurrence probability, the design level should be selected. For this reason the design should (i) evaluate sliding performance due to waves with larger recurrence intervals, and (ii) investigate overall sliding performance over the entire lifetime of the structure.

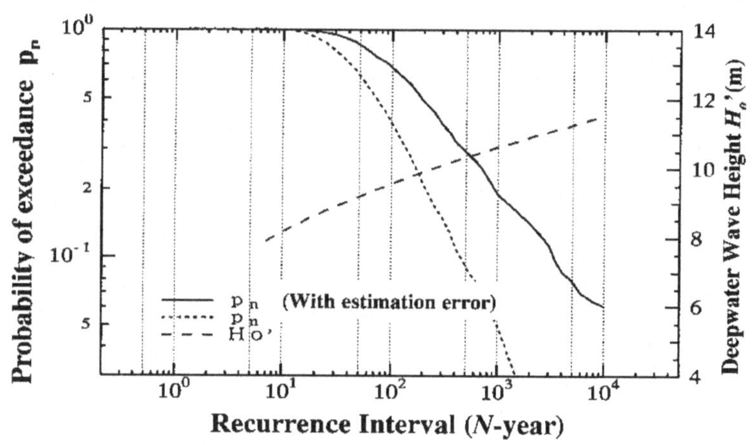

Figure 3 Probability of exceedance for a wave with various recurrence intervals over a 50-yr lifetime.

3.2.5. *Accumulated sliding distance over structure life-time*

Figure 4 shows the probability of exceedance of accumulated sliding distance over a 50-yr breakwater lifetime, where the mean value of the accumulated sliding distance, which we call the "expected sliding distance," is 30 cm. The probability of exceedance for a sliding distance of 1 m is 5% and 0.5% for 10 m. Note that the value of 30 cm corresponds to a 17% of probability of exceedance.

3.2.6. *Stability performance versus caisson width*

Figure 5 shows caisson sliding distance vs. caisson width for four design levels, where expected sliding distance for a 50-yr lifetime is also shown. When caisson width B = 22.1 m, the conventional sliding SF = 1.0, and the mean value of sliding distance for a 50-yr recurrence interval is 20 cm and the expected sliding distance is 81 cm. In contrast, for SF = 1.2 (B=26.5m), the sliding distance is very small, i.e., the mean value for a 50-yr recurrence interval wave is 1 cm and the expected sliding distance is 5 cm.

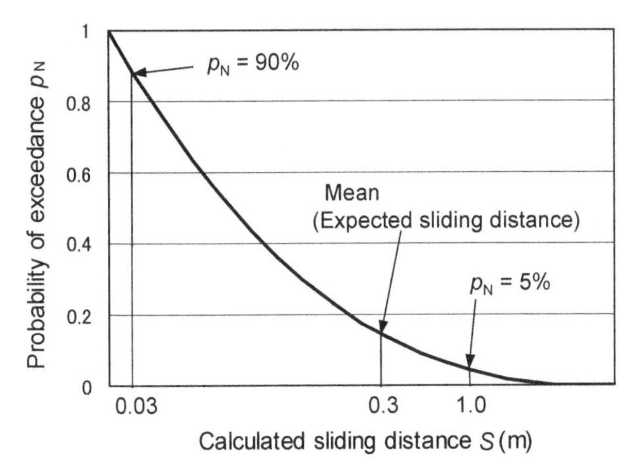

Figure 4 Probability of exceedance of accumulated sliding distance over a 50-yr lifetime.

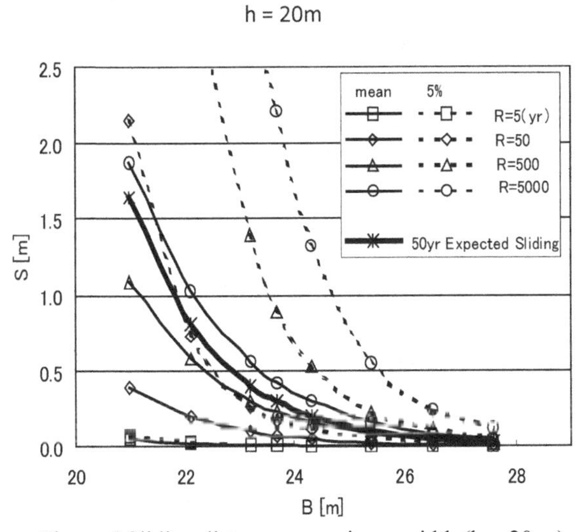

Figure 5 Sliding distances vs. caisson width (h = 20 m).

3.3. *Design Method Based on Stability Performance*

3.3.1. *Limit state design with a performance matrix*

Table 4 shows a so-called performance matrix first introduced by earthquake engineers. The vertical axis is the design level corresponding to waves with four different recurrence intervals, while the horizontal axis is the performance level defined by four limit states; namely, serviceability limit, repairable limit, ultimate limit, and collapse limit, corresponding to the extent of deformation. Serviceability limit and ultimate limit are defined in the current limit states design, whereas we added the other two limit states to more quantitatively describe the change of performance. That is, the collapse limit state is defined as extremely large sliding such that the breakwater falls off the rubble foundation, while the repairable limit state is deformation that is repaired relatively easily.

These limit states are defined by deformation, being the mean value of the sliding distance in this case. The values indicated here are so-called design criteria or allowable limits and are tentatively determined slightly conservatively, taking into account that the 5 % exceedance value is 3 or 4 times larger than the mean value. Letters A, B, and C in Table 4 denote the importance of a breakwater, i.e., A is critical, B is ordinary, and C lesser degree. For example, if a breakwater is classified as B, the necessary width of the caisson becomes less than 23.2 m for the sample breakwater.

Table 4 Performance matrix of a caisson.

Performance Level

Design Level / Limit states	Serviceability (3 cm)	Repairable (10cm)	Ultimate (30cm)	Collapse (100cm)
5-year	B	C		
50-year		B	C	
500-year	A		B	C
5000-year		A		B

3.3.2. *Design with lifetime sliding distance (expected sliding distance)*

The caisson width can also be determined considering the expected sliding distance obtained from the probability distribution of the accumulated sliding distance during a 50 yr lifetime. Table 5 shows the design criteria or allowable limit value of the expected sliding distance for breakwater classified as A, B, or C. For example, if the breakwater is classified B, the design criteria for the expected sliding distance width is 30 cm and the resultant caisson width is 23.68 m.

The value of 23.2 m determined from the limit state design with the performance matrix can be used as a design width. A width of 23.68 m can also be used as the final design value if lifetime performance is considered. The determined width is about 10% smaller than the conventional design value, i.e., the caisson width is reduced by clarifying its stability performance. For a breakwater in high importance category A, the necessary caisson width is 26.5 m to ensure that the expected sliding distance is less than 3 cm, while that for one in less importance category C is 21.2 m for an expected sliding distance of 100 cm.

Table 5 Design criteria using expected sliding distance.

Importance of Structure	A	B	C
Expected Sliding distance (cm)	3	30	100

3.3.3. *Deep water example*

Figure 6 shows the sliding distance versus caisson width for each design level in addition to the lifetime sliding distance. The water depth h is 30 m in this case, being quite deep compared to that in Fig. 5. Using the performance matrix (Table 4), the necessary caisson width for ordinary importance B is 22 m, while that determined by the expected sliding distance of 30 cm is 23.9 m; a value that corresponds to a sliding SF of 1.3, which exceeds a SF of 1.2 corresponding to a width of 22.1 m. Obviously the maximum wave height in deep water is not limited by water depth and therefore the necessary width becomes larger than the conventional design value. Accordingly, deformation-based reliability design is capable of effectively evaluating stability performance, with the resultant width closely corresponding to stability performance.

For a caisson width of 23.9 m in which the expected sliding distance is 30 cm, the sliding distance due to a wave with 500-year recurrence interval is 20 cm and that for one with a 50-year recurrence interval is 4.6 cm. The design criteria for expected sliding distance usually gives a larger width than that determined by the performance matrix; hence the expected sliding distance determines the necessary caisson width. This tendency is in fact intended when determining the design criteria for the performance matrix, although to confirm stability performance it is still necessary to check the caisson's sliding distance due to waves at each design level.

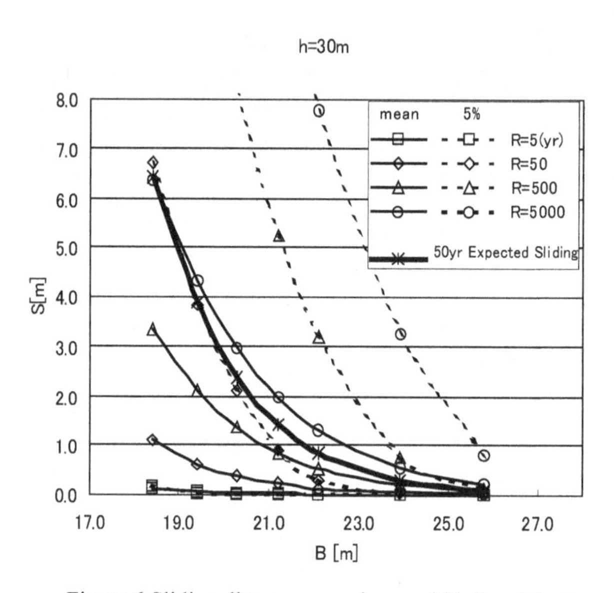

Figure 6 Sliding distance vs. caisson width (h = 30 m).

4. Deformation-Based Design for Armor Stones of Rubble Mound Breakwater

4.1. *Procedures to Evaluate Stability Performance*

4.1.1. *Deterministic value for a deepwater wave with recurrence interval*

Table 6 shows the flow of the calculation method of damage level of armor stones due to a deepwater wave with particular recurrence interval. After specifying the deepwater wave, the incident wave is calculated by the wave transformation method (Goda 1985, 1988). The van der Meer Formula is used to evaluate the damage level S of armor stones due to the incident wave. The damage level S is defined by the erosion area Ac around still-water level and nominal diameter of the stones D (van der Meer, 1988).

Table 6 Flowchart for calculating damage level.

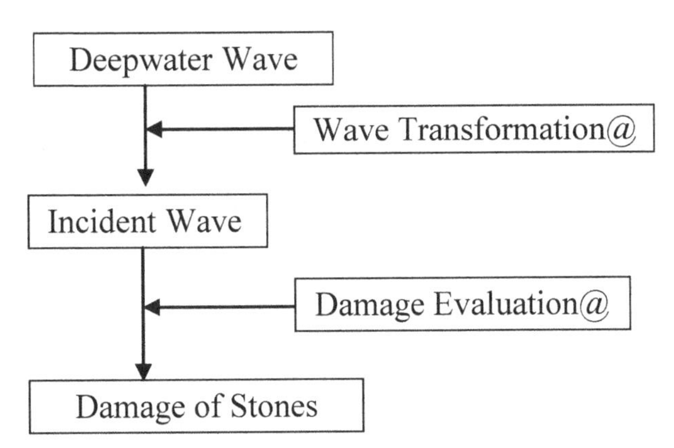

4.1.2. *Probabilistic value for a deepwater wave of a recurrence interval*

Even for a fixed deepwater wave condition, resultant damage of stones usually fluctuates due to the probabilistic nature of propagating waves and the response of the armor stones. To obtain the probabilistic damage level for a given deepwater wave, fluctuation of the items denoted by @ should be considered. Table 7 shows paramcters considered to reflect probabilistic nature in the present calculations and indicates bias of mean values and standard deviation (variance) of the probability distribution. The Monte Carlo simulation allows calculating the probability distribution of the damage level, with the calculation being repeated more than 2000 times from wave transformation to determination of the damage level for a fixed deepwater wave condition. From the probability distribution, the mean and 5% exceedance value are selected to represent the calculated distribution

Table 7 Estimation error for design parameters.

Design parameters	Bias of mean value	Variance
Wave transformation	0.	0.1
Size of stones	0.	0.1
Mass density of stones	0.	0.05
Deepwater wave	0.	0.1
Tidal level	0.	0.1

4.1.3. *Accumulated value during lifetime*

To obtain the accumulated damage level during breakwater lifetime (50 yr), one needs to consider the probabilistic nature of the deepwater wave and tidal level. The Weibull distribution with k = 2.0 is assumed as the extreme wave distribution with estimation error of 10% standard deviation. The tidal level is assumed as a triangle distribution between the L.W.L. and H.W.L. with error of 10% standard deviation.

A total of 50 deepwater waves are sampled and the damage level is evaluated by the Monte Carlo simulation using the procedure in Table 2. Total damage level due to the 50 deepwater waves is the accumulated damage level for a 50-yr lifetime. It should be noted that to calculate the accumulated damage level the equivalent number Nq of waves is necessary. It is assumed that the preceding damage level is caused by the equivalent number of waves Nq and that the damage level caused by the following storm is evaluated by the hypothetical number of waves (N-Nq). The probability distribution of the accumulated damage level is obtained by repeating the calculations more than 2000 times

4.2. *Example of Performance Evaluation*

4.2.1. *Design condition of a sample breakwater*

Stability performance of armor stones is illustrated here using a sample cross section of rubble mound breakwater as shown in Fig. 7. The conventional design condition of the breakwater is that the water depth h=10m, wave period T=10.1s, deepwater wave height (a 50-yr recurrence interval) Ho=4.76m, wave height at the breakwater $H_{1/3}$=4.65m (Number of waves N=1000). The designed mass of the stones M=11.7t when the armor slope is 1:2, the designed damage level S=2 and armor notional permeability factor P=0.4.

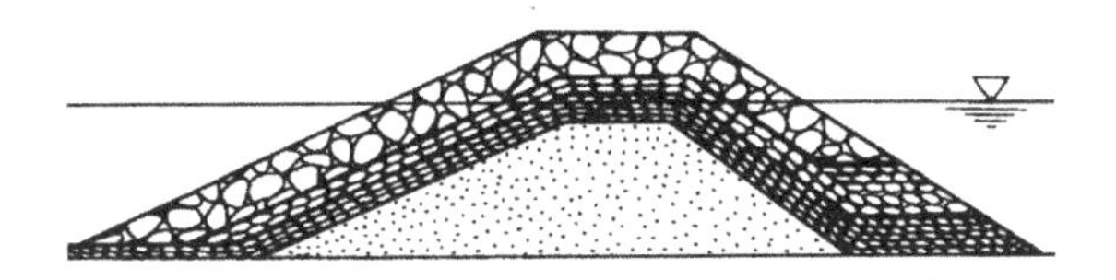

Figure 7 Cross section of a rubble mound breakwater for performance evaluation.

4.2.2. *Deepwater wave and damage level (deterministic value)*

Figure 8 shows damage level of armor stones of rubble mound breakwater produced by deepwater waves of different recurrence intervals, where the deterministic value of the damage level denoted by □ is 2 when the design wave with a 50-yr recurrence interval attacks, as designed.

Note that as the deepwater wave height increases, the damage level increases gradually; i.e., the damage level for a 500-year recurrence interval is 3.9, while the damage level for a 5-year recurrence interval is 0.9.

4.2.3. *Deepwater wave and damage level (probabilistic value)*

Figure 8 also shows the damage level due to deepwater waves obtained from the Monte Carlo simulation that included fluctuation of waves and block damage, where the mean (◇) and 5 % exceedance (↑) values of relative damage are indicated. Due to the probabilistic nature, i.e., the occurrence of larger incident wave height and larger damage, even the mean value of the relative damage is greater than the deterministic values. In fact, the 5% exceedance value is much larger the mean value. For example, for a wave with a 50-yr recurrence interval the mean value of the damage level is 2.9 and the 5% exceedance value is 8.7, whereas for a wave with a 500-yr recurrence interval the values are 5.5 and 16.0, respectively. Obviously then, the probabilistic nature must be considered.

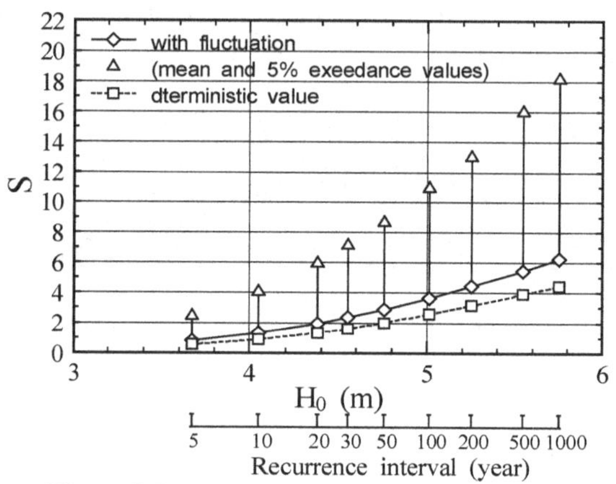

Figure 8 Deepwater wave height vs. damage level.

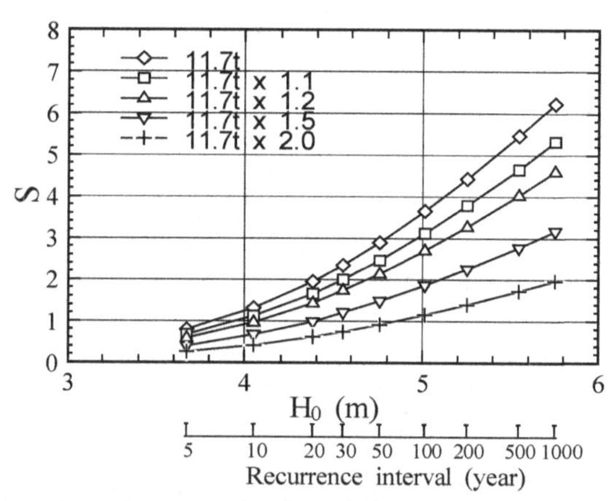

Figure 9 Damage level for different stone masses.

4.2.4. *Damage level for different stone masses*

Figure 9 shows the mean value of damage level for different stone masses. When the mass is 1.2 times the design mass, the mean value of damage level for a 50-yr recurrence interval is 2.13, 75% of the mean value for the design mass. When the mass is 1.5 times the design mass, the mean value of damage level is 1.47, a half of the mean value for the design mass.

4.3. *Performance Matrix for Armor Stones*

Table 8 shows a so-called performance matrix for armor stones. The vertical axis is the design level corresponding to waves with four different recurrence intervals, while the horizontal axis is the performance level defined by four limit states corresponding to the extent of deformation (damage). These limit states are defined by deformation, being the mean value of the relative damage in this case.

Table 8 Performance Matrix for Armor Stones.

Design Level	Performance Level			
	I	II	III	IV
5 (year)	B	C		
50		B	C	
500	A		B	C
5000		A		B

Degree I	**Serviceable**	**S=2**
Degree II	**Repairable**	**S=4**
Degree III	**Near Collapse**	**S=6**
Degree IV	**Collaspse**	**S=8**

Table 9 shows relation between the value of damage level and the actual extent of damage given by van der Meer(1988). For the case of 1:2 slope, the initial damage is when S=2 and the actual failure (exposure of the filter layer) is defined by S=8. In the performance matrix the serviceable limit is defined by S=2, Repairable S=4, Near collapse S=6 and collapse S=8, considering the definition in Table 9. For the structure of ordinary importance B, required stability performance is S=2 for a wave of 50-yr recurrence interval, and S=8 for 5000-yr recurrence interval in the performance matrix. If the sample breakwater is with ordinary importance the mass of 11.7t is enough to satisfy the performance matrix.

The values indicated here are so-called design criteria or allowable limits and are tentatively determined. Using the figure like Fig. 3 with Table 9 we can determine the necessary mass of the armor stones.

Table 9 Damage level S vs. Failure level.

Slope	Initial Damage	Intermediate Damage	Failure
1:1.5	2	3-5	8
1:2	2	4-6	8
1:3	2	6-9	12
1:4	3	8-12	17
1:6	3	8-12	17

4.4. *Lifetime Stability Performance*

Figure 10 shows the probability of exceedance of accumulated damage level over a 50-yr breakwater lifetime for different stone masses. The mean value of the accumulated damage level, which we call the "lifetime damage level or expected damage level," is 8.2 for the stones of the design mass. The probability of exceedance for an accumulated relative damage of 16.6 is 5%.

The value of the accumulated damage level of 8.2 is 4 times the design value and corresponds to the mean damage level for the wave of 5000-year recurrence interval. It can be said that the damage is accumulated significantly even by relatively small storms during the 50-year lifetime. However, this can be due to the characteristics of the van der Meer Formula and the current calculation procedure to accumulate the damage in this paper. It should be noted that the square root function of N gives the relatively large damage level compared with the experimental results. Some modification was suggested by van der Meer

(1988) and Melby and Kobayashi (1998) proposed a damage formula including the progress of the damage.

Note that the accumulated damage level is a hypothetical value not considering the repair within the lifetime. The repair should be made before the accumulated damage exceeds the repairable damage level.

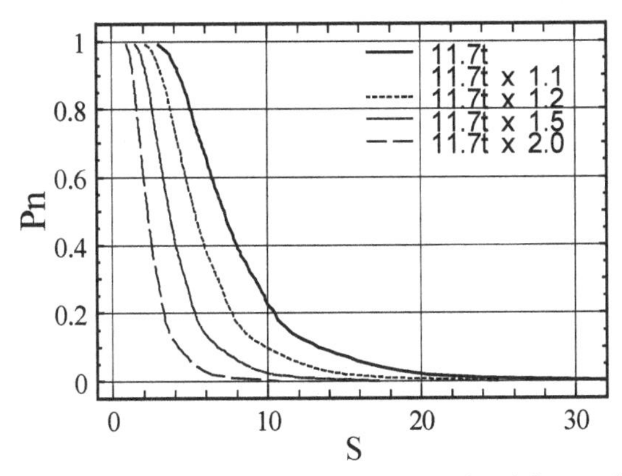

Figure 10 Probability of exceedance of accumulated damage level.

5. Concluding Remarks

5.1. *Scenario writing and accountability*

Nowadays it is crucial to obtain the understanding of the general public regarding the construction of coastal structures. Designs must incorporate accountability, and the performance of the facility must be explicitly and clearly explained. The best way for the citizens to understand the performance is to see what is actually happened when the storm attacks. To have better understanding from citizens, the performance should be described vividly like a scenario.

The actual failure is a prototype performance evaluation of the structure against the occurred storm, and the intensive investigation on the disaster usually done after the disaster is like a writing a scenario to describe what was happed by the storm. If such a scenario is made in the design stage, the stability performance can be understood very clearly by citizens. The performance design should include many scenarios for different occasions ie., different levels of storms. The design with many scenarios is actually the performance design, and therefore the performance design can be said as a scenario-based design.

5.2. *Performance design for coastal defenses*

The performance design should be extended to the design of coastal defenses for storm surges and tsunamis. The performance design for storm surge defenses was discussed (Takahashi et. al., 2004) after the disaster due to Typhoon 9918 in 1999. We are conducting a study to employ a performance design concept in the design of coastal defenses. Especially after the Indian Ocean Tsunami Disaster and the Hurricane Katrina Disaster, the importance of preparation for the worst case scenario—a case exceeding ordinary design levels—was pointed out (PIANC Marcom WG53 2010 and 2023).

Table 10 Performance matrix for tsunami defenses.

	Design tsunami	**Required Performance**
Level 1 Tsunami	Largest tsunami in modern times (return period: around **100** years)	**Disaster Prevention** • To protect human lives • To protect properties • To protect economic activities
Level 2 Tsunami	One of the largest tsunami in history (return period: around **1000** years)	**Disaster Mitigation** • To protect human lives • To reduce economic loss, especially by preventing the occurrence of severe secondary disasters and by enabling prompt recovery

Table 10 shows the measures to be taken in a worst-case scenario under the performance design which is considered after the Great East Japan Earthquake and Tsunami Disaster. The worst case is defined as a Level 2 tsunami assuming an occurrence probability of one every 1000 years, while a Level 1 tsunami is based on a conventional tsunami assuming a probability of one every 50 or 150 years. When planning for the conventional tsunami scenario, we aim to prevent the tsunami disaster. We aim to save lives, property, and the economy. On the other hand, the worst-case tsunami scenario considers disaster mitigation. The goal is to save lives, reduce damage to property, and prevent catastrophic damage to ensure early recovery.

5.3. *Future performance design*

Although the design technology for maritime structures has seen great advancements during the 20th century, the design frame is still unchanged. The design technology should be integrated to meet higher level of society's demand in this century. In addition to having scenarios related to stability and functional performance, the future performance design should include the scenarios on durability and environmental aspects including amenity, landscape, and ecology.

For the future performance design the technology to evaluate properly the performance should be further developed. Techniques for hydraulic model experiments using wave flumes and basins have significantly contributed to the development of maritime structure designs during the latter half of the 20th century. In fact, it is critical to conduct model experiments that evaluate not only wave forces but also resultant deformation in performance design. Since deformation of subsoil and breakage of structural members cannot be properly reproduced using small-scale experiments, large scale experiments are naturally paramount.

Numerical simulations are also important for investigating wave transformations and wave actions on structures including wave forces, especially by the introduction of direct simulation techniques (Isobe et al., 1999). Such simulations can explicitly show the process of wave propagation and action, which makes them quite suitable for performance design and use in the design process. Obviously then, both hydraulic model experiments and numerical simulations are important tools in performance design.

Acknowledgments

Sincere gratitude is extended to Professor Emeritus Y. Goda of Yokohama National Univ., Professor K. Tanimoto of Saitama Univ., and Professor T. Takayama of Kyoto Univ. for valuable comments on vertical breakwater studies. Discussions on the new design methodology with members of the working group within the Coastal Engineering Committee of JSCE were extremely beneficial in writing this paper; especially those with Professor T. Yasuda of Gifu Univ. and Professor S. Sato of Univ. of Tokyo.

References

Burcharth, H. F. (1993): Development of a partial safety factor system for the design of rubble mound breakwaters, PIANC PTII working group 12, subgroup F, Final Report, published by PIANC, Brussels.

Burcharth, H.F.(2001):Verification of overall safety factors in deterministic design of model tested breakwaters, Proc. International Workshop on Advanced Design of Maritime Structures in the 21st Century(ADMS21), Port and Harbour Research Institute, pp.15-27.

Goda, Y. (1973): A new method of wave pressure calculation for the design of composite breakwaters, Rept. of Port and Harbour Research Institute, Vol. 12, No. 3, pp. 31–70 (in Japanese).

Goda, Y. (1985): Random Seas and Design of Maritime Structures, Univ. of Tokyo Press, 323 p.

Goda, Y. (1988): On the methodology of selecting design wave height, Proc. 21st International Conference on Coastal Engineering, Spain, Malaga, pp. 899–913.

Goda, Y. and Takagi, H. (2000): A reliability design method for caisson breakwaters with optimal wave heights, Coastal Engineering Journal, 42(4), pp. 357–387.

Isobe, M, Takahashi, S., Yu, S. P., Sakakiyama, T., Fujima, K, Kawasaki, K., Jiang, Q., Akiyama, M., and Oyama, H. (1999): Interim report on development of numerical wave flume for maritime structure design, Proceeding of Civil Engineering in the Ocean, Vol. 15, J.S.C.E., pp. 321–326 (in Japanese).

Kimura, K., Mizuno, Y., and Hayashi, M. (1998): Wave force and stability of armor units for composite breakwaters, Proceedings 26th International Conference on Coastal Engineering, pp. 2193–2206.

Kobayashi, M., Terashi, M., and Takahashi, K. (1987): Bearing capacity of a rubble mound supporting a gravity structure, Rpt. of Port and Harbour Research Institute, Vol. 26, No. 5, pp. 215–252.

Melby, J. A. and Kobayasi, N. (1998): Damage progression on breakwaters, Proc. 26th International Conference on Coastal Engineering, pp. 1884–1897.

Oumeraci, H., Allsop, N. W. H., De Groot, M. B., Crouch, R. S., and Vrijling, J. K. (1999): Probabilistic design methods for vertical breakwaters, Proc. of Coastal Structures'99, pp. 585–594.

PIANC Marcom WG53(2010): Mitigation of Tsunami Disaster in Ports, Report No.112, PIANC, 111p.

PIANC Marcom WG53(2013): Tsunami Disasters in Ports due to the Great East Japan earthquake, 113p.

SEAOC (1995): Vision 2000-Performance-based seismic engineering of bridges.

Shimosako, K. and Takahashi, S. (1999): Application of reliability design method for coastal structures-expected sliding distance method of composite breakwaters, Proc. of Coastal Structures '99, pp. 363–371.

Shimosako, K, Masuda, S., and Takahashi, S. (2000): Effect of breakwater alignment on reliability design of composite breakwater using expected sliding distance, Proc. of Coastal Engineering, vol.47, JSCE, pp. 821–825, in Japanese.

Takahashi, S. (1996): Design of vertical breakwaters, Reference Document No. 34, Port and Harbour Research Institute, 85 p.

Takahashi, S., Hanzawa, M., Sato, K., Gomyo, M., Shimosako, K., Terauchi, K., Takayama, T., and Tanimoto, K. (1998): Lifetime damage estimation with a new stability formula for concrete blocks, Rpt. of Port and Harbor Research Institute, Vol. 37, No. 1, pp. 3–32 (in Japanese).

Takahashi, S. and Tsuda, M. (1998): Experimental and numerical evaluation on the dynamic response of a caisson wall subjected to breaking wave impulsive pressure, Proc. 26th International Conference on Coastal Engineering, ASCE, pp.1986-1999.

Takahashi, S., Shimosako, K., Kimura, K., and Suzuki, K. (2000): Typical failures of composite breakwaters in Japan, Proc. 27th International Conference on Coastal Engineering. ASCE. Pp.1899-1910.

Takahashi, S, Shimosako, K., and Hanzawa, M. (2001): Performance design for maritime structures and its application to vertical breakwaters, Proceedings of International Workshop on Advanced Design of Maritime Structures in the 21st Century (ADMS21), PHRI, pp.63-75.

Takahashi, S., Hanzawa, M., Sugiura,S., Shimosako, K., and Vander Meer J.W. (2003) Performance design of Maritime Structures and its Application to Armour Stones and Blocks of Breakwaters, Proc. of Coastal Structures 03, pp.14-26.

Takahashi, S., Kawai, H., Takayama,T., and Tomita, T.(2004) : Performance design concept for storm surge defenses, Proceedings of 29th International Conference on Coastal engineering, ASCE, pp.3074-3086.

Takayama, T., Suzuki, Y., Kawai, H., and Fujisaku, H. (1994): Approach to probabilistic design for a breakwater, Tech. Note Port and Harbor Research Institute, No. 785, pp.1–36, (in Japanese).

Takayama, T., Ikesue, S., and Shimosako, K. (2000): Effect of directional occurrence distribution of extreme waves on composite breakwater reliability in sliding failure, Proc. 27th International Conference on Coastal Engineering, ASCE, pp.1738-1750.

Tanimoto, K, Furukawa, K., and Hiroaki, N. (1996): Fluid resistant forces during sliding of breakwater upright section and estimation model of sliding under wave forces, Proc. of Coastal Engineering, JSCE, pp. 846–850, in Japanese.

Van der Meer, J. W. (1987): Stability of breakwater armor layers -Design formulae, Coastal Engineering, 11, pp. 219–239.

Van der Meer, J. W. (1988): Deterministic and probabilistic design of breakwater armor layers, Proc. American Society of Civil Engineers, J. of Waterways, Coastal and Ocean Engineering Division, 114, No. 1, pp. 66–80.

Vrijling, J. K., Vorrtman, H. G., Burcharth, H. F., and Sorensen, J. D. (1999): Design philosophy for a vertical breakwater, Proc. of Coastal Structures '99, pp. 631–635.

CHAPTER 4

AN EMPIRICAL APPROACH TO BEACH NOURISHMENT FORMULATION

Timothy W Kana
Haiqing Liu Kaczkowski
Steven B Traynum

Coastal Science & Engineering Inc
PO Box 8056, Columbia, SC, USA 29202-8056
E-mail: tkana@coastalscience.com

This chapter presents an empirical approach to beach nourishment formulation that is applicable to a wide range of sites with and without quality historical surveys. It outlines some analytical methods used by the authors in over 35 nourishment projects which help lead to a rational design. The empirical approach depends on site-specific knowledge of regional geomorphology and littoral profile geometry, some measure of decadal-scale shoreline change, and at least one detailed condition survey of the beach zone. The basic quantities of interest are unit volumes (i.e. the volume of sand contained in a unit length of beach between the foredune or backshore point of interest and some reference offshore contour) as a simple objective indicator of beach health which can be directly compared with volumetric erosion rates and nourishment fill densities. The focus of the chapter is on initial project planning—establishing a frame of reference and applicable boundaries, and developing conceptual geomorphic models of the site; and on project formulation—defining a healthy profile volume, calculating sand deficits and volume erosion rates, and formulating nourishment requirements for a defined design life. Example applications are presented for the general case and a site on the USA East Coast.

1. Introduction

Beach nourishment is the addition of quality sand from non-littoral sources for purposes of advancing the shoreline. It is an erosion solution increasingly embraced along developed coasts because it "... stands in contrast as the only engineered shore protection alternative that directly addresses the problem of a sand budget deficit."[1] In many jurisdictions, hard erosion-control structures are discouraged if not outright prohibited. Even where seawalls, revetments, and groins are permitted, beach nourishment is often a mandatory prerequisite for government approvals of coastal structures.

This chapter presents an empirical approach to beach nourishment formulation. During the past thirty years, the authors have designed and managed over 35 nourishment projects involving ~20 million cubic meters (m^3) in a wide range of settings, including the Carolinas and New York (USA), Kuwait, and several Caribbean beaches.

In evaluating the causes and rates of erosion at dozens of sites, from high energy coasts like Nags Head (North Carolina USA) to low-wave, high-tide range settings like Kuwait, we have found no universally applicable design method for beach nourishment. Each site is unique and subject to its own controlling coastal processes and suite of sediments. However, there are certain measurements and analyses which can improve the chance of successful beach restorations and help insure that projects perform as planned. We introduce some design techniques that have helped in our projects while drawing on proven analytical tools that should be part of the coastal engineer's daily practice.

1.1. *Motivation*

Sand is the lifeblood of beaches around the world. With sufficient sand in the littoral zone, a profile evolves under the action of waves which shape a beach into forms that are at once predictable to a degree, but a great deal variable from place to place and week to week. Winds and currents further modify the profile, shifting sand from the dry beach to the foredune, or carrying sand away to other areas along the coast.

During the past century, many of the great recreational beaches of the United States have been restored and maintained by adding sand, including Coney Island (New York), Miami Beach (Florida), and Venice Beach (California).[2] In South Carolina, the authors' home state, 102 kilometers (km) out of 160 km of developed beaches have been nourished since 1954.[3] A majority (~80 percent) of these localities have not only kept pace with erosion, but seen the shoreline advance significantly, leaving a wider beach and dune buffer during storms between buildings and the surf (Fig. 1).[4]

Wide beaches and high dunes are the essential elements of healthy sandy coasts. As property values continue to rise and demand remains strong for beach recreation,[2] sand will be needed along the coast. Beach nourishment is a soft-engineering measure to counteract erosion. When executed successfully, the outcome—more sand on a beach—should be no different than the effect of natural accretion.

1.2. *Topics Covered*

We offer in the following sections some empirical methods which help lead to rational design drawn from our constructed-project experience. The goal is to offer guidance that can be applied in nearly any setting, including sites with and without a historical database.

The focus herein is on preliminary design and project formulation rather than on permitting, economics, and implementation. Implicit is the assumption that quality sand closely matching the native beach is available and used as a borrow source. Comprehensive guidance for all aspects of beach nourishment design, including predictive modeling, is available in various sources (e.g.[5,6,7]).

We offer guidance for:

1. Initial Planning: (a) establishing a project frame of reference and applicable boundaries for analysis, (b) utilizing unit volumes—the basic measure of nourishment, and (c) developing conceptual geomorphic models of the site.
2. Project Formulation: (a) defining a healthy profile volume, (b) calculating sand deficits and volume erosion rates, and (c) formulating nourishment requirements for a defined design life.

Fig. 1. Myrtle Beach (SC USA) wet sand beach in March 1985 at low tide looking northeast (upper). Same locality in May 2014 after nourishment in 1986, 1989 (post-Hugo), 1997, and 2008 (lower). Beach fills have buried the seawall, created a protective dune, and maintained a dry-sand beach.

2. Initial Planning

2.1. *Frames of Reference*

Beach nourishment is a time-limited solution to coastal erosion. Like any engineered work, it will have a finite lifespan. The key question to answer is how long a project will provide benefits in the form of better storm protection, increased recreational area, or expanded habitat.

2.1.1. *Time Scales of Interest*

There have been a wide range of project longevities in the United States. Some sites, such as Hunting Island (SC), have required nourishment every four years or so, and are in worse condition today than before the first beach restoration.[3] By comparison, other sites, such as Coney Island (NY) have gone 20 years between beach fills and are in better condition nearly a century after the initial restoration[8] (L Bocomazo, USACE–New York District, pers. comm., February 2014). Differences in performance are closely tied to the background erosion rate, controlling sand transport processes at the site, and the length of the project.[6] Hunting Island, for example, has eroded at >8 meters per year (m/yr) for the past century, whereas Coney Island's background erosion rate is <1 m/yr (Sources: SC Office of Ocean and Coastal Resource Management, New York State Department of Environmental Conservation).

When erosion rates are higher than ~5 m/yr, periodic nourishment may not be able to keep up with the sand loss and, therefore, may not be the most effective and economic alternative. Solutions that combine nourishment with hard sand-retaining structures like groins and segmented breakwaters should be considered in areas of high sand loss.[1,6] Fortunately, much of the coast is eroding at moderate rates. Ref 9, in a study of the U.S. coastline, determined that most of the developed beaches were changing by <1 meter per year (m/yr) during the 20th century. Such hindcasting of average shoreline change provides a probabilistic projection into the future. Over 10–50 years, therefore, this equates to changes on the order of 10–50 m unless sea-level rise accelerates or some major avulsion occurs which locally modifies the rate of change.

In our experience, a practical longevity for individual projects is in the range 8-15 years for most sites. Shorter-lived projects are more difficult to amortize and generate popular support. Longer-duration projects (>20 years) become problematic and non-deterministic because projections of fill longevity have to be extrapolated beyond our ability to accurately predict future weather (i.e. winds and waves) and model sand transport at decadal scales.

2.1.2. *Cross-shore Scales*

Beach nourishment is fundamentally a volumetric measure. While the goal may be to widen the dry beach by some linear distance, this only occurs via the addition of a volume of sand. Further, the beach widening applies across some finite width of the littoral zone under the same processes that mold and shape a beach profile.

Sand placed at the seaward edge of the dry beach will quickly disperse across the profile until equilibrium slopes and surf-zone morphology associated with sediment sorting are achieved.[10,11] The cross-shore dimensions of concern are, therefore, related to the contours likely to change position over the time period of interest. At decadal scales under normal wave conditions (including periodic storms), most of the profile change occurs between the foredune and the local "depth of closure" (DOC).[12,13] This cross-shore dimension defines the active littoral zone and can be estimated for sites where good-quality historical profiles into deep water are available.

Profiles consist of distance-elevation pairs (x-z) referencing a common baseline (starting point) in "x" direction and vertical datum in "z" direction. If "i" is a point along a profile, the distance from the baseline can be written as x_i, and the corresponding elevation as z_i. If there is a suite of n profiles for a specific location for every x_i, a corresponding $z_{i,j}$ for the jth profile can be found via interpolation of raw data.* The mean elevation ($\bar{z}_i$) at the ith point for a total of n profiles can be computed by:

$$\bar{z}_i = \frac{1}{n}\left(\sum_{j=1}^{n} z_{i,j}\right) \qquad \text{(Eq. 1)}$$

**Raw field data rarely have measurements at the same point along the profile as prior surveys. We recommend the user interpolate the raw data from all surveys to uniform spacing in order to facilitate the simple computation of the standard deviation (σ_i).*

A standard deviation (σ_i) in elevation at this point can then be computed by:

$$\sigma_i = \sqrt{\frac{1}{n}\left(\sum_{j=1}^{n}\left(z_{i,j} - \bar{z}_i\right)^2\right)} \qquad \text{(Eq. 2)}$$

The standard deviation (σ_i) at each position along the profile normally approaches zero some distance offshore, providing an indicator that the elevation of the suite of profiles is not significantly different beyond some particular contour. Because field surveys involving bathymetric data typically introduce small errors, we assume closure for the period of record when σ_i remains constant and smaller than ~0.15 m for a given set of profiles in most cases.

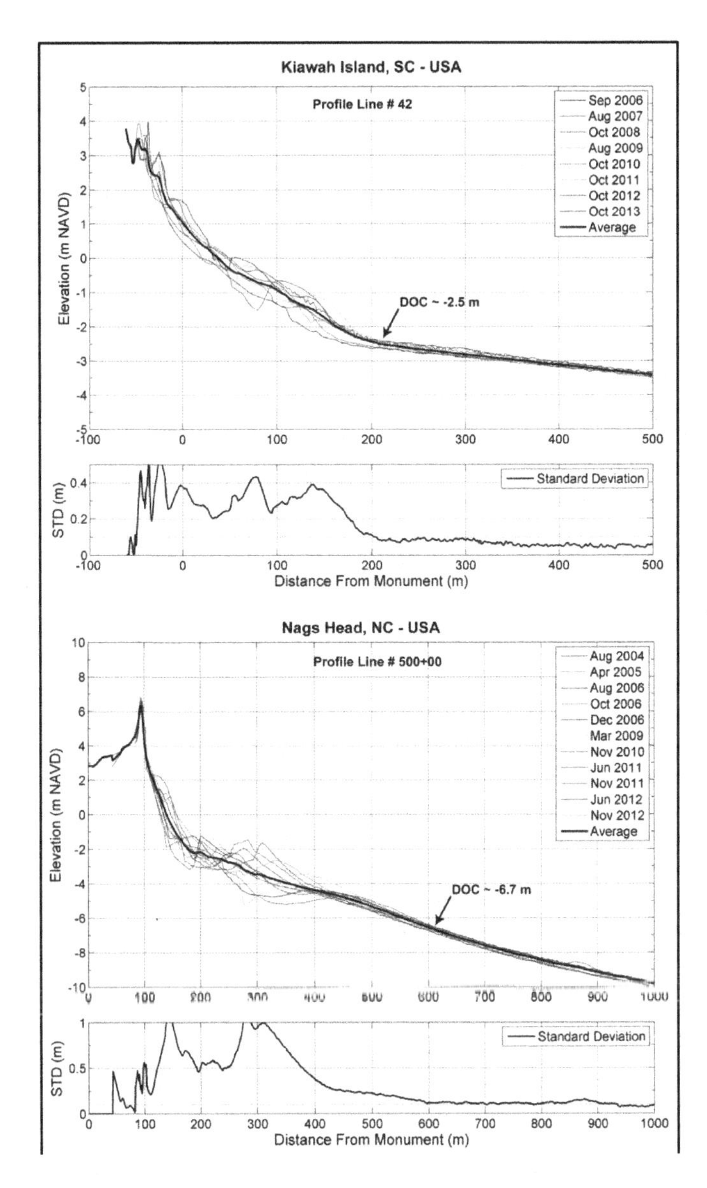

Fig. 2. Comparative profiles and plotted standard deviation (STD–σ) at two example transects extending from the foredune to deep water. Depth of closure (DOC) is based on σ approaching zero and remaining small for a significant distance offshore. DOC is shallower in low-wave settings.

DOC is related to incident wave energy and tide range[14] and can be estimated by empirical or analytical methods (Fig. 2). Ref 14 proposed the following empirical equation (Eq. 3) for estimating DOC (dn_l) based on the maximum "12-hour" wave height at a site. The 12-hour wave is the nearshore storm wave height (H_e) exceeded only 12 hours per year:

$$dn_l = 2.28\, H_e - 68.5\, (H_e^2)/gT_e^2 \qquad \text{(Eq. 3)}$$

where g is the acceleration due to gravity and T_e is the wave period associated with H_e.

High-energy beaches tend to have wider active littoral zones extending into deeper water compared with low-energy beaches. Accordingly, the volume needed to widen a high-energy beach is usually much greater than the volume for a sheltered beach.

Beach nourishment, to be effective and lasting, must replace annual losses along the coast in the zone between the foredune and DOC (Fig. 3). The typical elevation range for the littoral zone along most exposed shorelines is of the order 2–5 m above mean sea level (MSL) (backshore limit) and 2–10 m below MSL (seaward limit) based on field measurements.[12,15] As Fig. 3 implies, sand losses due to chronic erosion impact over a broad area in the littoral zone, but produce relatively small recessions at the foredune.

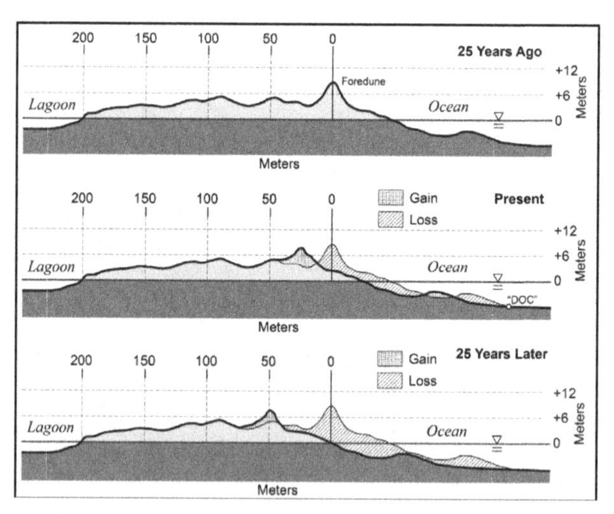

Fig. 3. Cross-sections through a gradually eroding barrier beach over a 50-year period, assuming dune recession averaging 1 m/yr. The "loss" section represents the littoral unit volume that would have to be replaced to maintain shoreline position at decadal scales.

2.1.3. *Longshore Scales*

The longshore frame of reference for nourishment varies according to the needs and objectives of the project, whether the goal is to address a limited segment or many kilometers of beach. Nevertheless, nourishment planning should consider the primary littoral cell within which a project lies.

Littoral cells are lengths of beach under the influence of similar controlling coastal processes.[16] Some classic examples include a pocket beach bounded by rocky headlands or a barrier island bounded by tidal inlets. Subcells are coastal segments nested within a primary littoral cell (Fig. 4). Offshore shoals, groins, and rocky outcrops along a beach modify sand-transport processes and mark secondary cell boundaries of interest.

The majority of nourishment projects are constructed to address erosion within some portion of a primary or secondary littoral cell. It is generally accepted that project longevity increases geometrically with length;[1,6] therefore, beach fills which encompass long reaches of shoreline tend to perform better. The majority of U.S. nourishment projects have length scales in the range 2–20 km.[7,8] The largest locally-funded beach nourishment project completed to date in the United States involved placement of 3.5 million m^3 of sand along 16 km of beach in Nags Head (NC).[17]

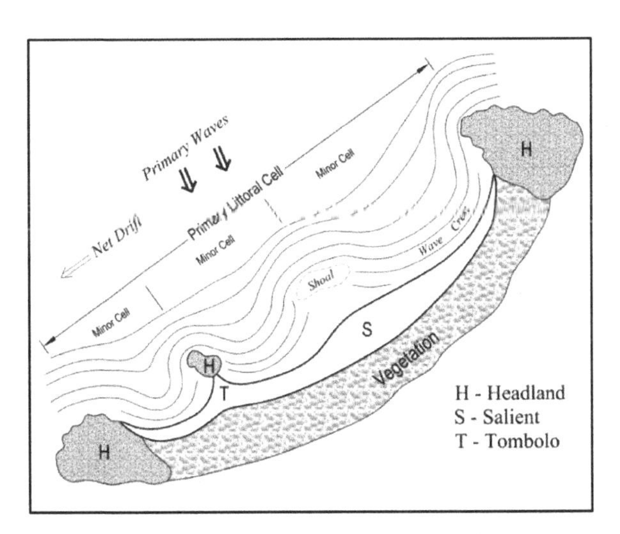

Fig. 4. Littoral cells are bounded segments of coast exposed to similar coastal processes. Islands, shoals, inlets, and coastal structures produce minor cells nested within a primary cell.

2.2. *Basic Units of Measurement*

The basic measure for beach nourishment is volume, usually given in cubic meters (or cubic yards). Often, popular accounts of projects report aggregate volumes (and costs) without regard to unit quantities or an explanation of the scale of a project with respect to the setting. For example, the impact and public perception of a 1 million m^3 project encompassing 2 km of beach is likely to be much different than a project involving the same quantity applied over 10 km.

For this reason, it is useful to consider beach nourishment in terms of "fill density" (ie – the volume placed over a unit length of shoreline). Common units for fill density (also referred to as "unit volume") are cubic meters per meter (m^3/m). The English equivalent is cubic yards per foot (cy/ft) which, conveniently, is almost exactly 0.4 times the metric value. In our example above, the 2-km project will have a fill density averaging 500 m^3/m (~200 cy/ft) of shoreline, whereas the 10-km project will average 100 m^3/m (~40 cy/ft).

Reporting volumes of nourishment in terms of fill density is useful in several respects. First, it quickly conveys a sense of scale in the cross-shore dimension for the project. Second, it provides a measure for direct comparison with other projects. And third, it provides a measure that can be directly related to volumetric erosion rates (Fig. 5).

In the United States, which continues to embrace English units, a fill density of 200 cy/ft (~500 m^3/m) would be considered large scale for most sites even if the project covers a short segment of beach. Beach fills <50 cy/ft (~125 m^3/m) are relatively small scale.

Regardless of absolute nourishment densities, the scale of a project should also be considered in relation to the setting and background erosion rate. Many beaches along low-energy shorelines have narrow profiles and evolve over a small elevation range relative to exposed ocean beaches. Their active profiles contain less volume and, therefore, nourishments involving small fill densities may have a relatively large and long-lasting impact. By comparison, a large fill density placed along a high-energy, rapidly eroding shoreline may yield relatively few years of erosion relief before the site returns to pre-nourishment conditions.

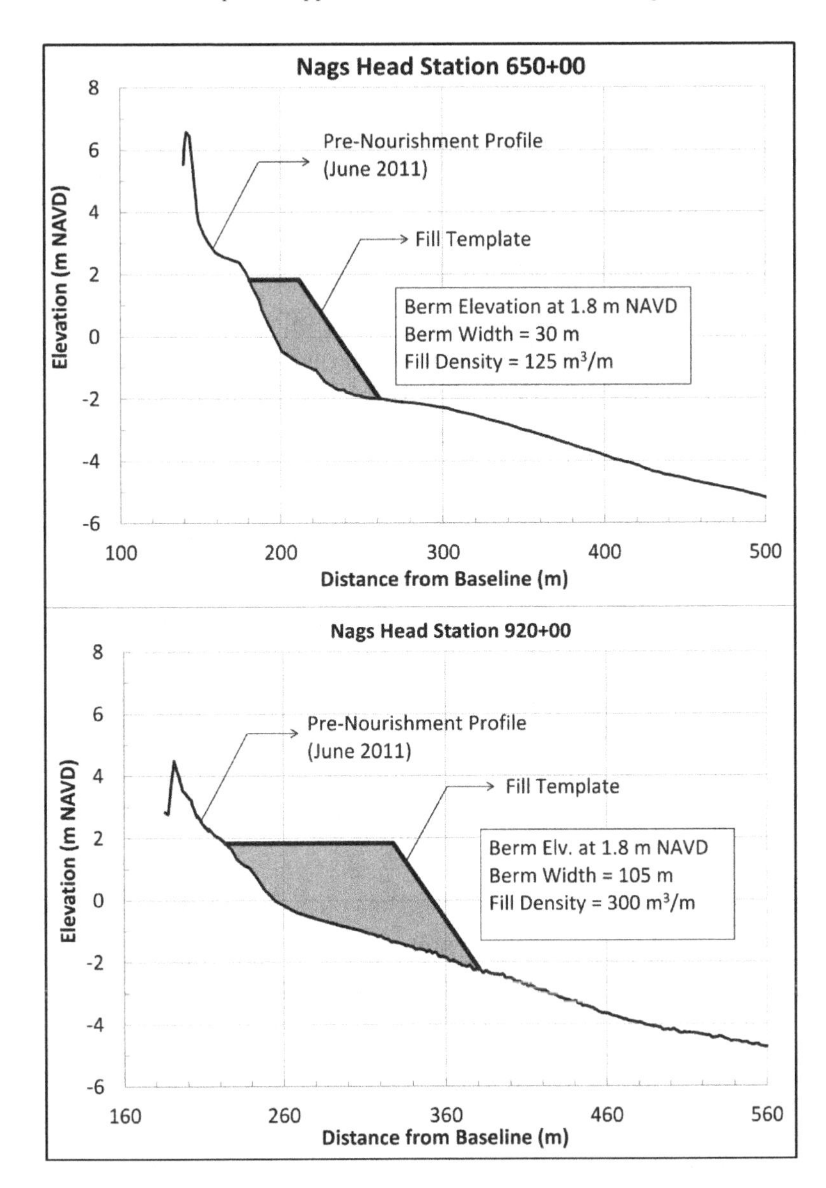

Fig. 5. Beach nourishment fill density—the volume per unit length of shoreline showing example design sections for a project in Nags Head (NC USA).[17] The larger fill density was applicable to areas experiencing shoreline recession rates >3 m/yr. The lower fill density was used along reaches eroding <1 m/yr. NAVD is North American Vertical Datum of 1988 which is a commonly used reference datum for surveys. NAVD is ~0.15 m above MSL along the U.S. East Coast.

Thus, the scale of a project is best considered with respect to three principal factors:

1) the underlying erosion rate for the site,
2) the dimensional scale of the receiving beach, and
3) the unit-width fill density.

By normalizing profile dimensions, erosion rates, and fill volumes using common "unit-width" measures, the critical design parameters can be directly compared.* For the rest of this paper, fill densities and unit-width volumes will be the principal units of measure for project formulation. Total volumes of beach sections are easily derived by extrapolation of unit volumes over applicable shoreline lengths.

**In recent years, improvements in surf-zone data collection techniques have led to an emphasis on surface models. Such 3D depictions of nearshore topography are highly instructive and provide detailed images of bar morphology and rhythmic beach features from which nearshore cell circulation and sand transport can be inferred (cf 18,19,20). Nevertheless, the standard measure for payment of beach nourishment quantities remains closely-spaced 2D profiles, which yield unit-width volumes or section volumes between profiles applying the average-end-area method. Because shore-perpendicular profiles remain the standard for payment and they are easily interpreted on construction drawings, we recommend their continued use for basic project design.*

2.3. *Conceptual Geomorphic Models*

Before developing a specific nourishment plan, it is useful to place the site in context and consider the regional controls on shoreline evolution. Conceptual geomorphic models provide a qualitative summary of site conditions and help focus plans for field data collection. As data are acquired, conceptual models can be refined to better indicate likely sand transport pathways for a project area. It is essential to know which way nourishment sand is likely to move after placement. Existing morphology provides clues.

Every site has unique characteristics—factors that make it unlikely any particular model (conceptual or numerical) will universally apply. However, there are recurring "signatures" of erosion which help place a particular beach in context and facilitate modeling and analyses.[21] At one end of the spectrum are long, linear beaches with little variation in sand transport under the influence of a steady wave energy and direction. At other localities, sand transport and shoreline morphology may be highly irregular because of seasonal wind patterns, offshore topography, tidal inlets, man-made structures, and other features that modify littoral processes.

The energy classifications of shorelines by Refs 22, 23, and 24 provide useful starting points for conceptual geomorphic models. Ref 22 divided the world's shorelines by tide range (micro <2 m, meso ~2-4 m, macro >4 m). Ref 23 described basic differences in barrier island morphology in terms of the relative importance of waves and tides. As tide range increases, or average incident wave energy decreases, there is a tendency for more frequent tidal inlets and sand bodies associated with tidal features. Increasing tide range also leads to differences in the subaerial beach with broader intertidal zones (i.e. "wet-sand" beach) relative to the "dry-sand" beach above the normal wave swash zone.

Sand spits, cuspate forelands, fillets at jetties, and tombolos in the lee of islands (*cf* Fig 4) are morphologic indicators of sand transport. These shoreline signatures help distinguish among beaches and provide broad measures of coastal processes acting on the site. Often, beach erosion problems are localized and may simply reflect some modification of shoreline processes due to changes in a nearby inlet, movement of offshore shoals, or some man-made alteration of the coast.[6] Figure 6 shows an example geomorphic model of a coastal segment in South Carolina (meso-tidal setting) bounded by major tidal inlets, subdivided by minor inlets, and further modified by natural washovers in one segment and groins in another subcell.

Conceptual geomorphic models provide an essential tool for beach nourishment design because they describe the overall setting and help the practicing engineer focus later analyses. An underlying goal should be to quantify the coastal processes, transport rates, and erosion quantities identified conceptually at the initial planning stage. Coastal geomorphology provides a framework.[25,26,27]

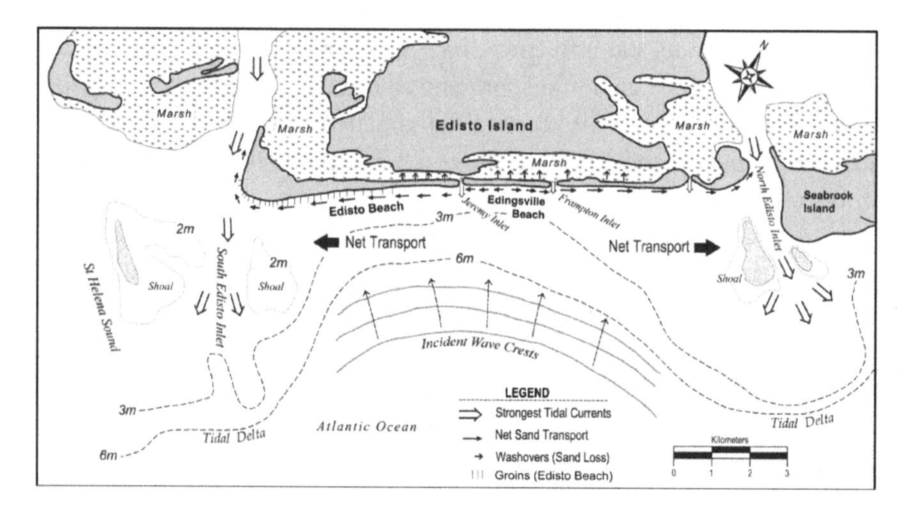

Fig. 6. Example geomorphic model of a segment of the South Carolina (USA) coastline illustrating the principal processes, bathymetry, and sand-transport directions for the area.

3. Project Formulation

3.1. *Shoreline Condition at Event Time Scales*

Geomorphic classifications of beaches based on the principal energy driving sediment are useful at decadal time scales. For shorter time scales, variations in wave energy and surge height in storms control shoreline position. The concept of a summer beach and a winter beach was introduced by Ref 28 and colleagues at Scripps Institute of Oceanography (USA) based on studies of onshore/offshore sediment transport along the California coast. During fair-weather waves in summer, the visible beach tends to build whereas, during periods of high waves in winter, sand moves offshore. This "beach cycle" can be seasonal but often it is controlled by the passage of storms[29] so the operating time scales are measured in weeks to months. In southern Asia, seasonal fluctuations in the profiles are controlled by the monsoon cycle.

The beach cycle produces characteristic changes in the profile including gentler slopes across the wave swash zone and vertical escarpments in the dry beach and foredune after storms. Fair-weather periods usually leave a wider dry beach and steeper wet-sand beach. Measurements of the visible beach alone will tend to be biased by the season or timing around storms. The two sets of profiles previously shown in Fig. 2 depicted conditions in different seasons and likely different numbers of days after storms. Yet, if we integrate the areas under the curves for each profile in the zone between the dune line and DOC (ie – zone where most change is occurring), we often find the differences in cross-sections are small. For randomly spaced profile data points (x-z), the area (A) under the profile can be calculated by:

$$A = \frac{1}{2}\sum_{i=2}^{n}\left(z_i + z_{i-1} - 2DOC\right)\left(x_i - x_{i-1}\right) \qquad \text{(Eq. 4)}$$

where n is the number of data points in a profile between two reference contours (in this case, between the dune line and DOC), i represents a random point, x_i and z_i are distance and elevation (respectively) from a fixed baseline and vertical datum, and DOC should be referenced to the same datum. Eq. (4) uses the "Midpoint Rule" when calculating area A. If data points are evenly spaced, "Simpson's Rule" (parabolic approximation) can be used. The area (A) under the profile is then:

$$A = \frac{b}{3}\sum_{i=2}^{n-1}\left(z_{i-1} + 4z_i + z_{i+1} - 6DOC\right) \qquad \text{(Eq. 5)}$$

where b is a constant and $b = x_i - x_{i-1}$.

Figure 7 shows the likely variation in area (A) to DOC versus the variation in position of a single subaerial contour normalized around the average for each value. Figure 7 illustrates how the full profile area (A) simply integrates the fluctuations in profile shape and yields a relatively constant value for each survey corresponding to coasts with negligible mean shoreline change. The contour position, by comparison, can be highly variable from survey to survey (Fig 7, middle and lower). Further, if we choose a different contour, the result may be the complete opposite with the erosion events causing seaward movement of certain underwater contours while some subaerial contours shift landward.

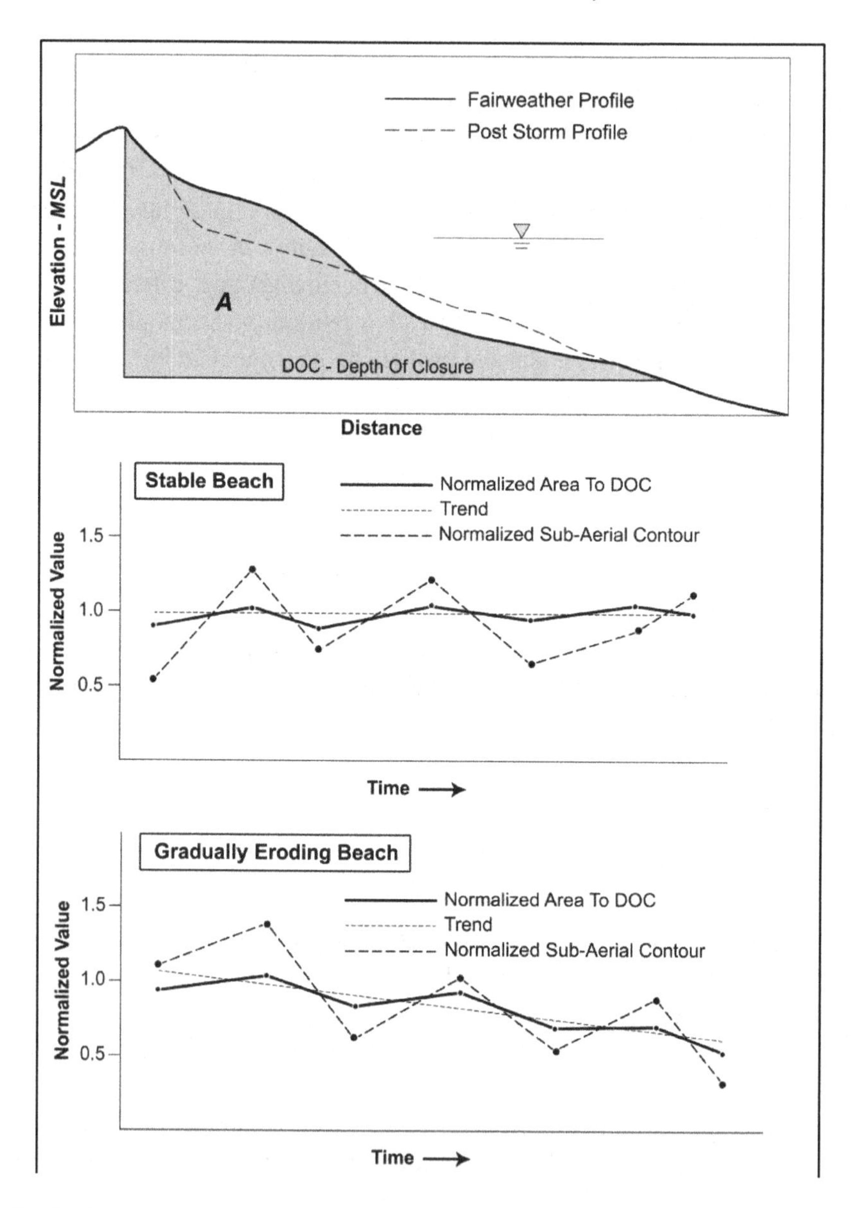

Fig. 7. Fair-weather and post-storm profile (upper) illustrating the idealized variation in cross-sectional area (A) to DOC versus the cross-shore position of a single subaerial contour, normalized around the average for each value. The middle profile represents a stable beach with negligible volume change. The lower profile represents a gradually eroding beach. Area A, when calculated to DOC, remains relatively constant for stable beaches because it integrates all the small-scale profile changes.

Profile areas computed to an arbitrary but fixed datum and encompassing the active littoral zone from the foredune to the estimated DOC integrate all the small-scale perturbations across a beach. This eliminates most of the seasonal bias in surveys and yields a relatively pure measure of profile condition.[30] Obviously, if the site in question is losing sand, "*A*" will diminish over time with respect to a fixed point landward of the foredune (*cf* Fig. 7, lower).

3.2. *Profile Volumes as a Measure of Beach Condition*

It is a simple matter to convert a two-dimensional cross-sectional area to an equivalent volumetric measure. If *A* is extrapolated over a 1-m length of beach, the result is a unit volume (*V*) in m^3/m (English units are typically cy/ft—conveniently 1.0 cy/ft equals ~2.5 m^3/m). We now have an absolute volume reflecting the condition of the beach at one profile along the coast. Unit volume is a fundamental unit of measure for the littoral zone from which erosion analyses and nourishment formulations can be derived. Use of absolute unit volumes may not have the mathematic elegance of non-dimensional data, but they provide a site-specific quantity for direct application. Our empirical design approach seeks measurements that are site-specific for purposes of determining specific nourishment volumes to place on a beach. Another reason to utilize absolute profile volumes at a site is they give the practicing engineer a sense of scale and proportion, which becomes important over time as more experience with beach erosion and nourishment is gained.

Table 1 provides typical unit volumes for a number of beaches the authors have evaluated. We also list some corresponding parameters that affect the dimensions of the active littoral zone. As the data illustrate, unit volumes fluctuate over a wide range from place to place. This should not be surprising considering that the profile geometry and morphology will vary with tide range, wave energy, DOC, sediment size distribution, and in some cases, underlying lithified sediments which fix the bottom substrate (particularly in tropical environments dominated by calcium-carbonate sediments).

Table 1. Example profile unit volumes from foredune to indicated contour for sites evaluated by the authors.

Locality (State USA)	Mean Tide[a] Range (m)	Wave Ht[b] (m)	Mean Grain Size[c] (mm)	~DOC[d] (m below MSL)	Calc Depth (m below MSL)	~Slope[e] Swash Zone	~Unit Volume[f] (m^3/m)	Nourishment Density (m^3/m)	Beach Type[g]	Reference
Bridgehampton (NY)	0.9	1.2	0.45	8.2-9.1	3.0	0.070	560	210	Barred Beach	31,32
Bridgehampton (NY)	0.9	1.2	0.45	8.2-9.1	5.5	0.070	1315	210	Barred Beach	31,32
Fire Island (NY)	1.0	1.1	0.43	7.3-9.1	3.0	0.065	410	N/A	Barred Beach	31,33
Fire Island (NY)	1.0	1.1	0.43	7.3-9.1	6.1	0.065	1500	N/A	Barred Beach	31,33
Fire Island (NY)	1.0	1.1	0.43	7.3-9.1	9.1	0.065	3350	N/A	Barred Beach	31,33
Nags Head (NC)	1.0	1.4	0.41	8.5	7.3	0.060	2150	220	Barred Beach[h]	17
Nags Head (NC)	1.0	1.4	0.41	8.5	5.8	0.060	1375	220	Barred Beach	17
Nags Head (NC)	1.0	1.4	0.41	8.5	3.7	0.060	625	220	Barred Beach	17
Nags Head (NC)	1.0	1.4	0.41	8.5	1.8	0.060	315	220	Barred Beach	17
Bogue Banks (NC)	1.1	1.0	0.33	6.1	3.4	0.050	500	180	Barred Beach[i]	34
Myrtle Beach (SC)	1.5	0.8	0.28	3.6-4.6	4.6	0.045	610	255	Non-barred Beach[j]	35
Myrtle Beach (SC)	1.5	0.8	0.28	3.6-4.6	1.7	0.045	200	255	Non-barred Beach	35
Isle of Palms (SC)	1.5	0.6	0.25	4.0	3.0	0.035	750	N/A	Non-barred Beach	36
Kiawah Island (SC)	1.6	0.5	0.20	3.0	3.0	0.030	700	N/A	Non-barred Beach	37
Edisto Beach (SC)	1.7	0.5	0.58	3.8	3	0.050	350	120	Non-barred Beach	3,38
Hunting Island (SC)	1.9	0.4	0.22	3.8	3.8	0.035	650	835	Non-barred Beach[k]	39
Kuwait City	3.0	0.15	0.38	1.5-2.5	2.5	0.100	275	N/A	Perched Beach[l]	40
St Lucia -SW Coast	0.2	0.15	0.30	2.0	2.0	0.125	150	13	Pocket Beach	41
Rose Hall, Jamaica	0.5	0.1	0.45	1.5	1.5	0.125	100	N/A	Protected by Reef	42

Footnotes:

[a]Source: National Oceanic and Atmospheric Administration (NOAA).

[b]Typical inshore average breaking wave height based on various sources including cited reports and authors' observations.

[c]Typical subaerial beach between the toe of dune and low tide wading depth. Deeper water samples are typically much finer.

[d]DOC—Depth of Closure at "decadal" time scales based on cited reports and comparative profiles obtained by the authors.

[e]Typical slope (inverse of rise over run) for the active swash zone between ~+1 m and –1 m MSL.

[f]Original data collected in English units (cy/ft), converted to metric equivalents (1 cy/ft ~=2.5 m^3/m), and rounded for typical pre-nourishment profile at site.

[g]Barred beaches maintaining a longshore bar composed of finer grained sediments than the subaerial beach and separated from the swash zone by a well defined trough;

[g]Non-barred beaches with sediments that are typically fine sand (<0.25 mm) across the entire profile.

[h]Average nourishment density for one beach fill (2011) over a length of 16.1 km.

[i]Cumulative average nourishment density based on multiple beach fills (2001 to 2007) over a length of 28.9 km.

[j]Cumulative average nourishment density based on four beach fills (1986 to 2009) over a length of 14.9 km.

[k]Cumulative average nourishment density based on eight beach fills (1968 to 2006) over a length of 4.8 km.

[l]Broad lithified platform (wave cut terrace) underlies beach at ~1.5 to 2.5 m below MSL.

The data in Table 1 also include some example unit volume quantities for portions of the littoral profile. While these partial volumes will not reflect equilibrium unit volumes to DOC, they provide useful measures. For example, in the United States, the Federal Emergency Management Agency (FEMA)[43] recommends criteria for the unit volume in protective dunes seaward of development. FEMA prescribes the "540 rule" which is the cross-sectional area in square feet (ft^2) contained above the 100-year return-period, maximum still-water level seaward of the foredune crest. The equivalent unit volume for a 540-ft^2 cross-section is 20 cy/ft (50 m^3/m). Sites meeting this criterion are generally considered to have adequate protection for backshore development during the 100-year storm event. Beach communities with high protective dunes tend to sustain lower damages in storms and, therefore, file fewer insurance claims.

Many sites have a long history of subaerial profile surveys which terminate in shallow water. This partly reflects the ease of land-based surveys relative to profiles encompassing the surf zone and inshore area to the DOC. Profile volumes for subaerial beaches will contain a fraction of the equilibrium volume to DOC. Nevertheless, as data are acquired at a site and extended into deeper water with more accuracy over time, early surveys to low water or some wading depth yield volumes that can be compared with volumes to DOC. In so doing, ratios can be developed between the subaerial volume and the volume to closure.[35] These ratios become important for rapid storm assessments where initial surveys after an event are limited to land-based data collection.

3.3. *Calculating Sand Deficits and Volume Erosion Rates*

While a single profile to DOC can provide a fundamental measure of beach condition, a suite of profiles is needed to quantify sand deficits and erosion rates. Commonly, networks of profiles* are established for a segment of coast with sufficient density to distinguish longshore variations. We have found that a practical spacing for shoreline erosion analyses along exposed ocean coasts is 500–1,000 ft (~150–300 m).

Such spacing generally captures the variation in beach morphology while allowing efficient field data collection. More closely-spaced profiles are needed near inlets to account for delta shoals and rapidly changing wave exposures.[6,44]

**Profile networks can also serve as a basis for payment in many nourishment projects. In the U.S., the standard for pre- and post-nourishment surveys is a profile spacing of 100 ft (~30 m). Payments are based on before and after dredging surveys computing the differences in cross section applied over the distance between profiles using the average-end-area method.*

Normally, profiles are referenced to a baseline extending more or less parallel to the shoreline. Distances alongshore and profile azimuths reference the baseline. For practical reasons, baselines often follow primary roads paralleling the beach, or some reference infrastructure or survey monuments. But for quantitative comparisons of beach condition along a segment of coast, individual profile volumes should reference a primary morphologic feature such as the dune crest or seaward vegetation line (Fig. 8). Such features are natural indicators of the average maximum limit of wave runup at a site. Dune crests and vegetation lines are generally more stable features than the mean high waterline, berm crest, or some other contour in the active surf zone.

Rectified aerial imagery provides a useful tool for establishing a relatively consistent backshore starting point for unit volume calculations. Vegetation lines can be digitized by closely spaced points and a smoothed line constructed via floating-point averages. Where the smoothed line intersects surveyed profiles, a "starting distance" for unit volume calculations relative to the baseline is determined. With consistent* starting points established for each profile the variation in unit volumes alongshore is minimized.

**Vegetation lines can be highly variable, especially in areas of accretion where incipient dunes are forming. For this reason, it may be preferable to reference starting points at some profiles to a projection of some primary dune feature which is detectable along the rest of the coast.*

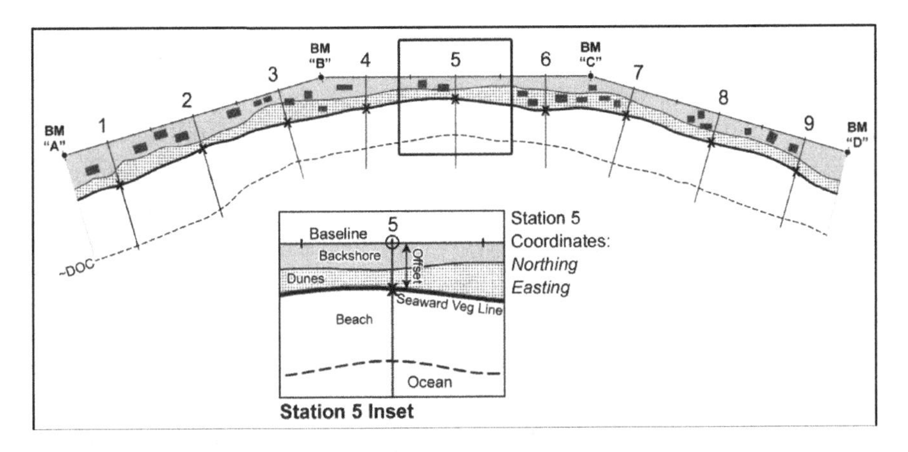

Fig. 8. Typical project layout showing a survey baseline with fixed control points (BM – benchmarks), profile lines extending offshore beyond the local DOC, and delineation of a key morphologic feature such as a dune crest or a seaward vegetation line. Profile unit volumes referencing the dune crest or existing buildings provide an objective measure of beach condition from section to section.

The seaward limit for unit volume calculations is ideally related to the local DOC. For practical reasons, it is useful to select a fixed contour and adopt a single value for calculations in a given segment of coast. This provides common reference boundaries for comparing profile volumes from one line to another (Fig. 9).

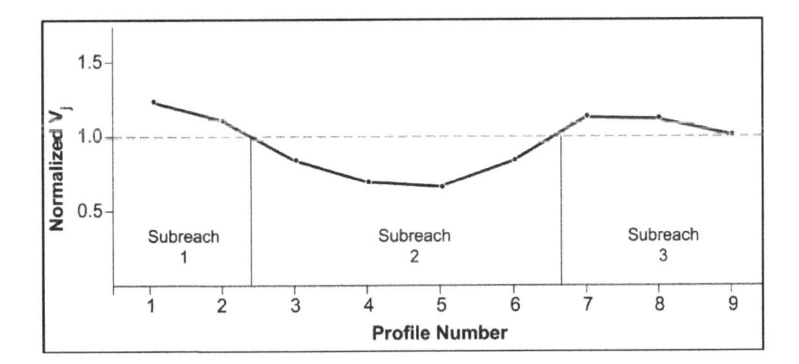

Fig. 9. Normalized unit volume (V_j/V_{mean} for the jth profile) for a suite of profiles (see Fig. 8) measured between a common backshore feature (dune crest/dune escarpment/seaward vegetation line) and a fixed offshore contour (ideally ~DOC). Profiles along Subreach 2 contain ~70–80 percent of the mean volume and have a sand deficit relative to survey Subreaches 1 and 3.

3.3.1. *Absolute vs Relative Quantities*

Unit volumes for a suite of profiles collected at the same time provide a measure of the relative health of a beach at each line in the absence of historical data. They can also be extrapolated to provide a fundamental measure (the control volume) for applications in sediment budgets.[45]

It is not uncommon to observe systematic variations in unit volume alongshore. These variations can be compared with the average volume for a reach, thereby distinguishing zones with less sand or more sand than average. The volumes can be normalized against the average to quickly indicate the relative surplus or deficit. However, absolute unit volume differences tend to be most instructive because they provide a measure that can be related directly to a nourishment volume density.

We find that it is useful to pool groups of profiles alongshore and compute averages for subreaches. This reduces the variability of individual lines while establishing broader trends for discrete reaches. Figure 10 illustrates a result for the generic cases shown in Figs. 8 and 9. The profile volumes for subreaches show a clear trend of lower volumes along Subreach 2.

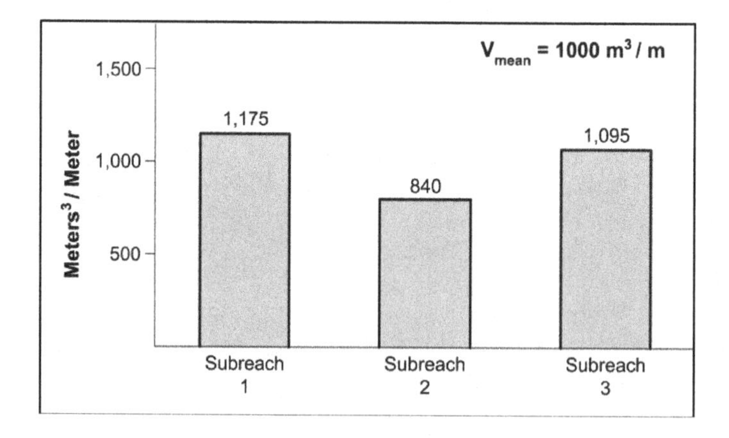

Fig. 10. Variation in average profile volume (V) by subreach for the generic cases illustrated in Figs. 8 and 9; in this case, assuming the mean $V = 1{,}000$ m³/m. At a glance, Subreach 2 has a deficit of ~160 m³/m compared with the average for the suite of profiles. Visual inspections in the field should provide morphologic indicators of these variations by reach.

In this case, we have assumed the average V is 1,000 m³/m for the suite of profiles to a referenced depth contour. Profiles in Subreach 2 contain ~85 percent of the mean unit volume. The adjacent subreaches contain ~9–18 percent greater volume. The implication is that Subreach 2 is less healthy (in terms of beach dimensions) and has a sand deficit. A site visit and closer inspection of those profiles would likely confirm whether the deficit is in the visible beach (e.g. dune escarpment, narrower berm) or underwater (e.g. missing longshore bar). The difference could also be associated with previous manipulation of the profile such as emergency dune scraping—a technique that can maintain a dune line while the littoral profile loses volume. Absolute quantities in Fig. 10 provide an immediate approximation of sand deficits. Subreach 2, for example, contains ~160 m³/m less sand than average for the suite of profiles.

3.3.2. *Referencing Unit Volumes to Development*

Profile unit volumes can also be referenced to existing development. Rather than using a common morphologic feature along the backshore, it is sometimes useful to compute volumes seaward of buildings or shore protection structures. In these cases, variations in unit volumes will reflect man made conditions. However, their utility is in establishing the relative health of the beach and absolute quantities of sand seaward of structures. This provides an objective indicator of properties with better or worse protection within a community along the beachfront. Figure 11 provides a result for a site in North Carolina where the setback of buildings varies. At a glance, it is apparent that properties in the middle subreaches have much less beach/dune volume than adjacent reaches. Regardless of the cause of these differences or any systematic erosional trends, deficit quantities are readily estimated from a single set of profiles that can be obtained in an initial shoreline assessment.

We find it useful to compare sites lacking historical surveys with similar settings where more comprehensive data are available. If the sites have similar exposures and sediment quality, the site with history will provide examples of "healthy*" beach volumes.

**Healthy beaches are considered to contain sufficient sand to withstand the range of seasonal profile changes without adverse impact to the foredune. A probabilistic measure of health can be related to the volume contained in the dune above various return-period storm-tide levels.*[43,46]

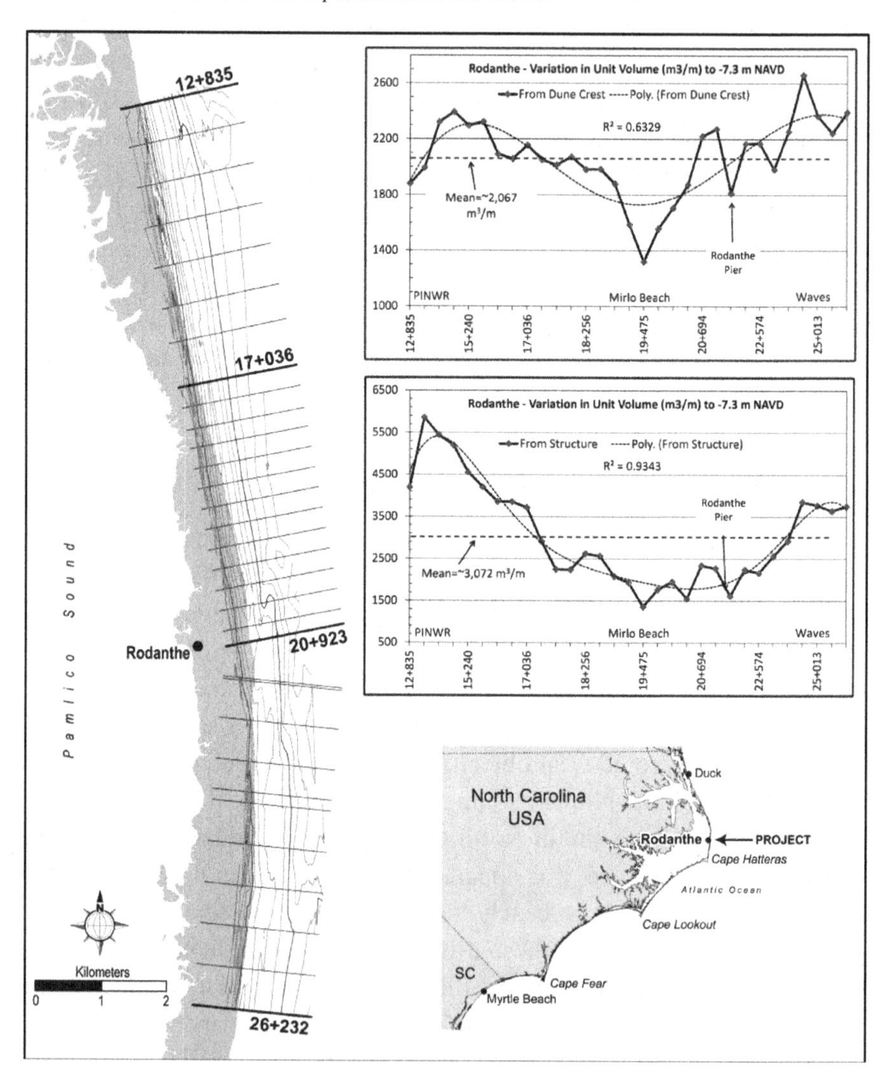

Fig. 11. Variation in actual unit volume to ~DOC for a 13-km ocean shoreline in Rodanthe (NC USA) with respect to the dune crest/seaward vegetation line (upper) and existing buildings (lower). Field data are rarely smooth due to variations in profile geometry, sediment-size distributions, or artificial manipulation. Nevertheless, the sections exhibiting a sand deficit are easily distinguished.

3.3.3. *Estimating Volume Erosion Rates*

Ideally, a site considered for nourishment will have quality historical profiles extending beyond the DOC. By reoccupying the historical survey lines and updating the profiles, present conditions can be compared with prior conditions. The profiles are overlaid in a common coordinate system and differences in cross sectional areas are computed. We prefer using fixed starting distances and a common depth limit (cutoff) so unit volumes by date (to common boundaries) as well as the unit volume change can be evaluated. Assuming a site is erosional, the starting distance determined for existing conditions during the initial design assessment will fall landward of prior vegetation lines or dune crests. Earlier surveys will, therefore, likely contain higher unit volumes unless the profile geometry has significantly changed.

For sites lacking quality historical profiles, volume erosion rates must be estimated by extrapolation of linear shoreline change data. In many jurisdictions official erosion rates are established based on analysis of historical maps and aerial photographs. The quality of these data varies, but there are distinct advantages in using official rates when available. In cases where linear rates for different time periods are available, the results reflecting conditions closer to the present time are generally better indicators of near-future trends. Longer periods tend to yield lower erosion rates because of the effect of averaging. However, shoreline data predating aerial photography is often suspect because of inherent limitations of survey technology and positioning a century or more ago, and errors associated with digitizing paper map products.[47]

For nourishment planning, linear erosion rates must be converted to an equivalent volumetric erosion rate. This conversion can be made two ways. First, by displacing existing profiles in the cross-shore direction—the volume erosion rate is the change in cross sectional area extrapolated over a unit length of shoreline.[7] The second method is easier and only requires an estimate of the height of the profile over which normal change occurs (*cf 48,49*). The zone of normal change is generally assumed to extend from the average dry beach elevation to DOC. It can be shown that a vertical column of sand 1 m^2 in area (Fig. 12) contains the same volume to DOC as a wedge of sand following the contours of a profile. Thus, the unit volume of sand to DOC (V) in 1 m² of dry beach is:

$$V = h \quad \text{(Eq. 6)}$$

where h is profile height (m) from the dry-beach level to DOC and V is in m³/m.

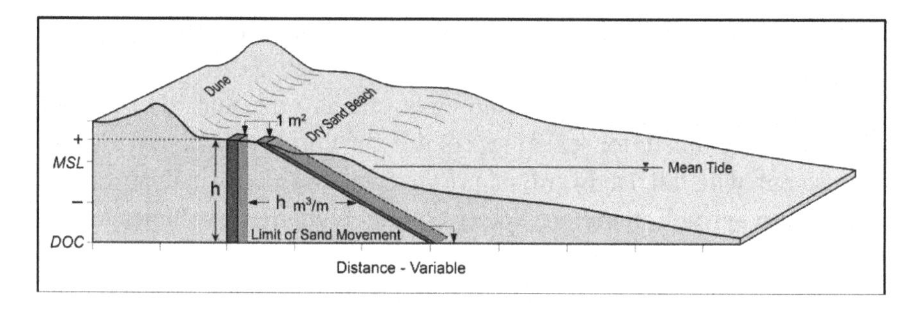

Fig. 12. The unit volume of sand to closure along a profile in 1 m² of beach area is equal to a vertical column of sand with height h as illustrated.

If the linear erosion rate at a site is L_e in m/yr, the equivalent unit volumetric erosion rate (V_e) can be computed by:

$$V_e = L_e \bullet h \quad \text{(Eq. 7)}$$

where V_e is in m³/m/yr. If L_e is in English units of ft/yr and h is in ft, V_e in cy/ft/yr is then:

$$V_e = \frac{1}{27} L_e \bullet h \quad \text{(Eq. 8)}$$

With an equivalent volume erosion rate for a site, the cumulative erosion for various planning periods can be estimated. We commonly project erosion for 5-year, 10-year, and 20-year periods. The extrapolations for longer periods become problematic where historical data are limited.

Other factors affect nourishment longevity, particularly project length and sediment quality.[6] However, for initial planning the "advance" nourishment part of the project formulation should be directly related to historical changes in some way. Adjustments in the form of safety factors, fill diffusion rates, and so on, can be applied at final design.

3.4. *Formulating Nourishment Volumes*

Standard practice in the U.S. is to establish a design profile for beach restoration which may or may not incorporate a protective dune (or storm berm) and some minimum dry-beach width.[1,6] This profile is the target minimum cross-section to be maintained over the life of the project. Nourishment construction associated with the design profile is considered a "first cost" or initial volume of federal projects.[7] Some extra volume is typically added to the design profile in anticipation of future losses. The extra volume is referred to as "advance nourishment" in U.S. federal projects. Thus, the standard formulation in the U.S. considers an initial base volume plus an advance volume with advance nourishment quantities linked to some defined period of time.

Our empirical approach to nourishment design similarly has two components. However, the initial volume is not predicated on a particular dune dimension or beach width, but rather the site-specific profile deficit volume relative to a healthy unit volume. Fill configurations can be in many forms but paramount is they augment the sand supply on the beach and raise unit volumes to some minimum quantity.

The emphasis is on sand placement in the active beach zone such that the added volume is exposed to waves. Nourishment profiles, upon placement, should include broad dry beach zones that are readily overtopped by waves. Narrow berms or berms situated well above the normal beach level will tend to leave scarps and require more time to adjust.

While it may be counterintuitive to place sand in the most energetic part of the beach, this has the advantage of producing a more natural profile soon after construction. If the fill berm is designed for an average elevation which matches nearby healthy dry beach elevations; it will overtop in minor events and sand will redistribute across the profile. Overtopped berms will promote onshore sand transport by wave swash and lead to a natural shift of sand toward the foredune. In areas with energetic sea breezes, onshore winds also shift nourishment sand across the berm and facilitate dune growth.

The decision whether to incorporate a protective dune or confine nourishment to the active beach zone depends on many factors including the pre-project backshore condition. The unit volumes needed for enhanced dunes tend to be dwarfed by the deficit volume across a typical profile. For this reason, it may not be worth extra expense to fill carefully around existing structures such as house piles and dune walkovers. Emphasis should be on restoring the active beach in most cases. Another advantage of nourishing primarily in the active beach zone is backshore access corridors for the public can sometimes be maintained during construction (Fig. 13).

Fig. 13. Ground photos showing backshore corridor for public access during construction. During nourishment operations, it is sometimes useful to set berm elevations slightly below the normal dry-beach elevation and limit work to the seaward part of the beach. The constructed berm will be subject to overtopping during minor storm events but will adjust more rapidly to a natural beach configuration.

3.4.1. *Preliminary Design by Reaches*

We typically formulate a preliminary design for a limited number of reaches using average values for the deficit and advanced nourishment quantities. The reaches should be selected based on observed trends in unit volumes and erosion rates among groups of profiles, as well as other factors such as average setbacks (exposures) of buildings, sizes of dunes, recreational access (i.e. demand for a segment of beach), or political boundaries. Use of broad averages has the advantage of reducing the variability of individual profiles and providing easily interpreted designs.

Figure 14 provides an example of a preliminary beach-fill plan for the hypothetical beach previously illustrated in Fig. 7, highlighting the deficit volumes and advance volumes using averages by reach. For the example, we assume the underlying erosion rate averages 5 m^3/m/yr to DOC in all subreaches. A "ten-year" advance nourishment volume would, therefore, be at least 50 m^3/m added to the deficit volume. We also assume, in this case, the mean unit volume for the given set of profiles constitutes the "minimum healthy volume" from which specific deficit volumes can be computed at each profile.

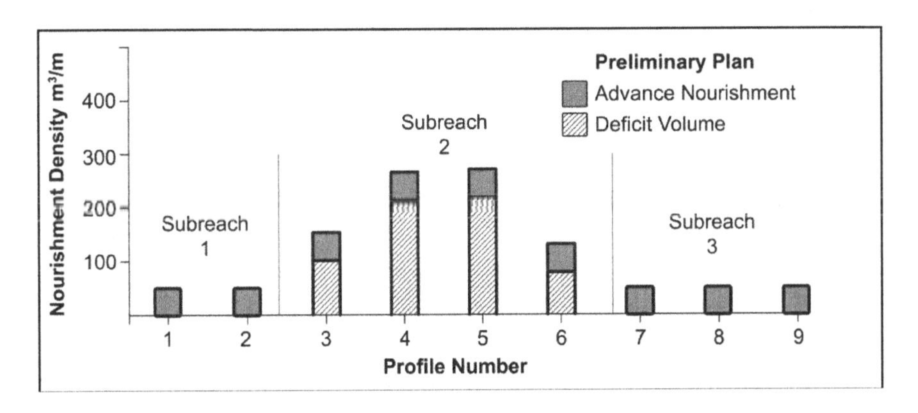

Fig. 14. Preliminary nourishment plan for the hypothetical beach referenced in Fig. 7. A "ten-year" beach fill involves two components: (1) deficit volume with respect to the mean unit volume which varies from line to line in Subreach 2 by an average of ~160 m^3/m and (2) advance volume which is ten times the background erosion rate (assumed = 5 m^3/m/yr). Often along highly eroded beaches, the deficit volume needed to restore a minimum healthy beach dwarfs the advance nourishment volume.

3.4.2. *Final Design Requirements*

The preliminary design provides the coastal community with an estimate of the scale of nourishment required for a specific longevity and guidance for the necessary funds to implement the project. The community may seek public support and apply for construction permits after receiving the preliminary design. Final design must account for transitions in fill density from reach to reach, tapers at the project ends, and the cross-shore configuration of the fill. Importantly, final design must adjust for changing conditions along the oceanfront between the time of the preliminary design and the final design. While final design entails considerable work, it generally must be accomplished within the scales and tolerances allowed under permits for construction.

An advantage of our empirical approach utilizing average unit volumes by reach is its flexibility for the final design configuration. It is relatively easy to hold reach volumes constant (to satisfy the permitted quantities) while modifying the configuration of sand placement to fit conditions close to the time of construction. But the key point here is that no amount of refinement during final design can overcome a poorly formulated preliminary design.

Although there have been successful examples using numerical models in coastal engineering, the complicated nature of shoreline evolutions hinders the application of computer models in nourishment design. Simulation of shoreline change requires a combination of hydrodynamic, sediment transport and morphological models. The model setup and calibration demand great time and effort, and therefore, numerical modeling is normally not the best approach at the preliminary design stage of a project. However, once a basic project formulation has been completed, numerical models can assist in final design efforts and focus on refinement of the nourishment plan.

The authors have used the USACE-approved coastal engineering models such as GENESIS[50] and SBEACH[51] to facilitate the final design for Nags Head (NC USA)[17] and other sites. Computer models allow testing of fill diffusion and dispersion under various fill configurations, a check on the overall project performance as well as performance of individual sections of beach after nourishment, and a basis for modifying the design to increase the longevity of the project.[6] The modeling results can be used to identify the potential occurrence of erosion hot spots and optimize the nourishment design so the adverse effects of such hot spots are minimized. The results can also be used to evaluate potential short-term, storm impacts on the beach and further optimize the design.

The final section of this chapter presents a case application of our empirical method for nourishment design.

4. Example Application — Bogue Banks North Carolina USA

Bogue Banks (NC USA) is a 25-mile-long* (40-km), south-facing barrier island bounded by Beaufort Inlet and Cape Lookout to the east and Bogue Inlet to the west (Fig. 15). It has been positionally stable for centuries with low rates of shoreline change and no breach inlets. However, between 1995 and 1999, five hurricanes in quick succession impacted the island and caused repeated recession of the dunes, some of which exceed 30-ft (10-m) heights above the normal beach level. Since no hard shore-protection structures are allowed in this jurisdiction; individual property owners resorted to yearly sand-scraping and frequent replacement of dune walkovers.

**English units with metric equivalents are used in this case example, consistent with the original data collection, project formulation, and post-project monitoring.*

Bogue Banks encompasses five towns and one state park—from west to east: Emerald Isle (EI), Indian Beach/Salter Path (IB/SP), Pine Knoll Shores (PKS), Atlantic Beach (AB), and Fort Macon State Park (FMSP). Prior to 2000, only Atlantic Beach had received measureable nourishment (1986 and 1994), which occurred in the form of federal dredge material disposal associated with a nearby navigation project.

The authors assisted each community in formulating a beach restoration plan which initially involved ~26 km and over 3.6 million m^3 using sand from offshore borrow areas.

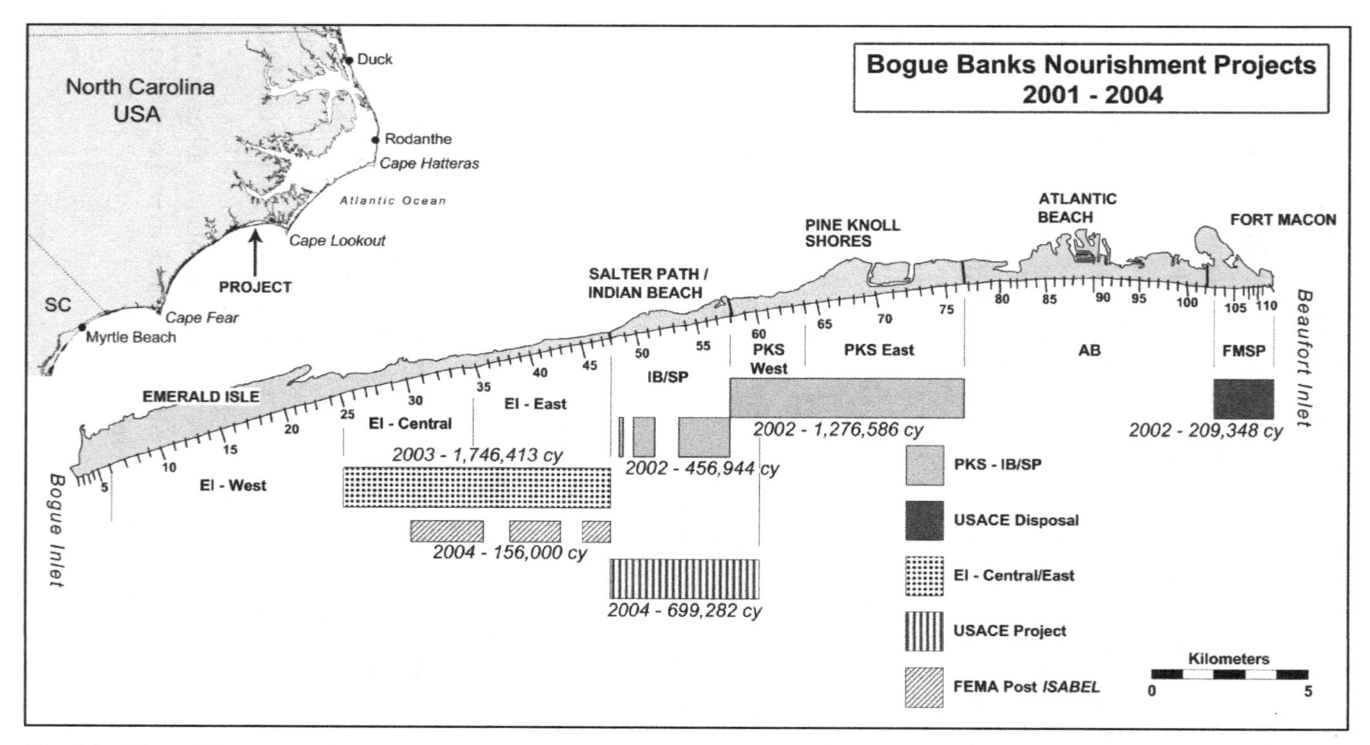

Fig. 15. Map of Bogue Banks (NC USA) showing profile lines and subreaches used in preliminary design for a series of beach nourishment projects designed by the authors. Volumes are in cubic yards (cy). 1 cy ≈ 0.7645 m³

4.1. *Initial Erosion Assessments and Preliminary Design*

Prior to 1999, there were no beach surveys to DOC over the majority of the island for comparison. However, the state of North Carolina publishes official erosion rate maps based on aerial photo interpretation. Nearly the entire island has eroded at rates between 0.5 m/yr and 1.0 m/yr in recent decades.[52]

To develop a preliminary nourishment design, the authors set up a baseline along the backshore and surveyed profiles from the baseline across the surf zone and outer bar at ~300-m spacing. The profiles were accomplished via rod and level using swimmers (a data collection technique that remains cost effective for projects in the developing world) and therefore terminated in intermediate depths [approximately −13 ft MSL (−4 m)] somewhat short of DOC which was estimated to be 15–20 ft (4.6–6.1 m) below mean tide level for the setting.* Precision RTK-GPS measurement systems and satellite data acquisition were just being released to the public around that time.

**Bogue Banks is midway between Duck (NC), where quality profiles over several decades have documented DOC at (~)−7.5–9 m MSL,[12] and Myrtle Beach (SC), where DOC is estimated to be (~)−3.6–4.6 m MSL.[35] Its south-facing beach is sheltered from predominant winds out of the northeast.*

With no comparative profile data, our design team developed unit-width volumes for standard cross-shore boundaries between the foredune and an offshore depth of (~)−11 ft (~3.4 m) MSL. The offshore boundary generally encompassed the longshore bar situated 400–800 ft (~120–240 m) offshore. Unit volumes by station were averaged by reach and subreach using political boundaries as principal dividing lines for reaches (Fig. 16). Profile volumes demonstrated differences among reaches, notably much lower volumes along EI, IB/SP, and PKS with respect to AB, which apparently retained some sand from earlier nourishment (dredge material disposal).

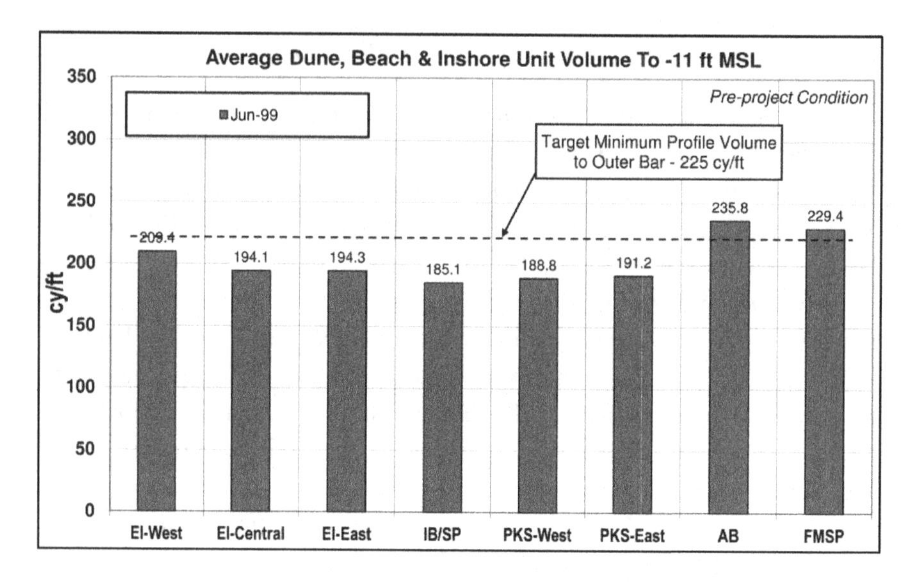

Fig. 16. Average profile volume by reach between the foredune and −11 ft (~3.4 m MSL) along Bogue Banks (NC USA) based on an initial site condition survey in June 1999. Only Atlantic Beach (AB) had been nourished prior to the survey. [Note: 1 cy/ft ≈ 2.5 m^3/m]

As the initial beach nourishment formulation and project costs were under review by the community, Hurricane *Floyd* impacted the area (18 September 1999), causing extensive damage and further dune recession. However, property damage was observed to be minimal along Atlantic Beach.[53] The principal difference along AB was its significantly higher profile volumes. Based on this favorable post-storm response, the AB beach condition (as measured by average profile volume to −11 ft (~3.4 m) MSL became the target minimum volume for other communities along Bogue Banks. For the profile boundaries used, the minimum "healthy" volume was calculated to be 225 cy/ft (~560 m^3/m).

The difference in unit volumes between AB and every other reach established rational deficit volumes. Final project formulation then combined the reach deficit with an advance nourishment volume linked to the average annual erosion rate, converting decadal average dune recession to an equivalent volumetric rate and extrapolating for a ten-year project longevity. Consistent with standard engineering practice for other infrastructure works, the advance volumes were increased by a "safety factor" of ~1.5 to account for uncertainty in the formulation. Final design efforts further refined the site-specific nourishment templates for numerous subreaches.[54]

4.2. *Project Implementation and Monitoring*

During the subsequent four years (2000–2003), individual nourishment projects were funded and implemented by each community and their performance tracked against the preliminary design and target minimum profile volumes. Summary graphs such as Fig. 17 provided each town on Bogue Banks with easily interpreted performance measures. Similar to the Dutch approach,[55] Bogue Banks now has a detailed record of its sand volumes in the littoral zone each year from which decisions can be made objectively to place additional sand as needed.

As of 2012, over 9 million m^3* have been added along Bogue Banks since the preliminary design work of 1999.[56] The reference unit-width volumes remain the primary benchmark for renourishment decisions. Importantly, the condition of the oceanfront remains improved beyond its condition of 1999 as evidenced by growth of a new foredune line and few permits issued for emergency sand scraping or reconstruction of dune walkovers.

**Approximately 25 percent of the total has been placed along Atlantic Beach as part of ongoing federal dredge disposal operations.*

5. Summary and Conclusions

The empirical approach to nourishment design outlined herein offers a rational method for initial planning and project formulation. It applies unit-volumes—a basic measure of beach condition—for purposes of determining the "physical health" of a site, calculating sand deficits, converting linear erosion rates to equivalent volume change rates, and selecting rational fill densities for a desired project longevity. The method is site specific and dependent on at least one high-quality survey of the active littoral zone to the normal limit of measurable volume change.

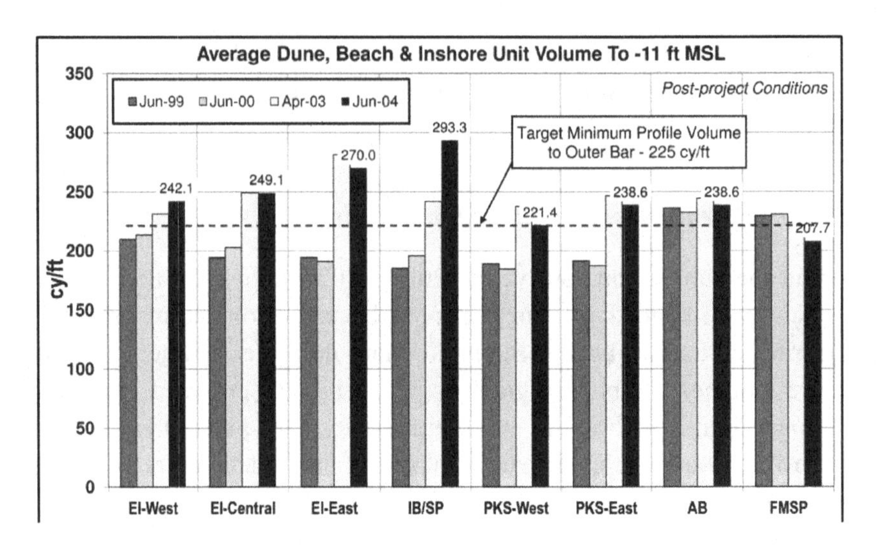

Fig. 17. Increase in profile volumes by reach following nourishment along Bogue Banks (NC USA) between 2001 and 2004. [See Fig. 15 for dates and volumes placed by reach.] These average reach volumes are tracked against a target minimum profile volume for the sites using defined reference contours. [Note: 1 cy/ft ≈ 2.5 m³/m]

We have utilized the empirical method for beach design in numerous projects and find that it offers a defensible basis and provides a readily understood formulation for developing community support, estimating realistic project costs, and receiving government approvals of the plan.

Beach nourishment requirements are highly variable along with site conditions such that no single formulation or design method is universally applicable. In our experience, the relatively few, or no, prior beach nourishments at most sites mean there will be high uncertainty of outcomes. A goal of our empirical method is to reduce the uncertainty. Quality beach surveys encompassing the entire littoral zone at a site are the single most important set of measurements for design. In the absence of detailed coastal process data (winds, waves, tides, and sediment quality data), unit profile volumes provide an absolute measure of beach condition; in effect, integrating all the small-scale, temporal perturbations of the littoral zone which develop in accordance with incident waves and sediment transport.

Preliminary design using the analytical methods described in this chapter is only one element of nourishment planning. Considerable effort is required for sound final design, including sediment testing, confirmation of borrow sources, and simulation of fill diffusion and dispersion. Following construction of a project, beach surveys should be performed periodically so that the basic quantities of concern—unit profile volume changes, volumetric erosion rates, and nourishment volume remaining—can be compared with pre-project conditions, and project performance can be objectively determined.

The practicing coastal engineer should seek to evaluate as many sites as possible, develop an understanding of the time and spatial scales of coastal erosion, then track completed projects carefully so as to continually improve future designs.

Acknowledgments

The preliminary design methods presented herein have evolved over the past 30 years during evaluation of coastal erosion at dozens of sites in a wide range of settings. The authors thank numerous municipalities for their support and commitment to not only maintaining their beaches, but funding ongoing beach survey programs which have been a key to successful projects, including the communities of Myrtle Beach (SC), Isle of Palms (SC), Edisto Beach (SC), Kiawah Island (SC), Seabrook Island (SC), Nags Head (NC), Emerald Isle (NC), Pine Knoll Shores (NC), and Southampton (NY). Additional research funding to the authors has been provided by the US Army Corps of Engineers, South Carolina Sea Grant Consortium, SC Office of Ocean and Coastal Resource Management (OCRM), and SC Department of Parks Recreation and Tourism (PRT).

We thank Prof Young C. Kim for the invitation to include a chapter in this volume. Trey Hair and Diana Sangster prepared the graphics and manuscript.

References

1. NRC, *Beach Nourishment and Protection.* National Academy Press, Washington, DC (1995).
2. JR Houston and RG Dean, Beach nourishment provides a legacy for future generations, *Shore & Beach.* **81**(3), 3-30 (2013).

3. TW Kana, A brief history of beach nourishment in South Carolina, Shore & Beach, **80**(4), pp 1-13 (2012).
4. JB London, CS Dyckman, JS Allen, CC St John, IL Wood, and SR Jackson, An assessment of shoreline management options along the South Carolina coast. Final Report to SC Office of Ocean and Coastal Resource Management; prepared by Strom Thurmond Institute for Government and Public Affairs, Clemson University (SC) (2009).
5. CUR. *Manual on Artificial Beach Nourishment.* Rept. No. 130, Center for Civil Engineering Research, Codes and Specifications (CUR), Rijkswaterstaat, Delft Hydraulics Laboratory, Gouda, The Netherlands (1987).
6. RG Dean, *Beach Nourishment: Theory and Practice* (World Scientific, NJ, 2002)
7. USACE. *Coastal Engineering Manual: Coastal Project Planning and Design – Part V.* US Army Corps of Engineers, EM 1110-2-1100 (2008).
8. CERC, *Shore Protection Manual.* 4th Edition, US Army Corps of Engineers, Coastal Engineering Research Center; US Government Printing Office (1984).
9. R Dolan, S Trossbach, and M Buckley, New shoreline erosion data for the mid-Atlantic coast, *Journal of Coastal Research.* **6**(2), 471-478 (1990).
10. RG Dean, Equilibrium beach profiles: characteristics and applications, *Journal of Coastal Research.* **7**(1), pp 53-84 (1991).
11. PD Komar, *Beach Processes and Sedimentation.* Second Edition, Prentice-Hall, Inc, Simon & Schuster (1998).
12. WA Birkemeier, Field data on seaward limit of profile change, *Journal of Waterway Port, Coastal and Ocean Engineering.* **III**(3), 598-602 (1985).
13. NC Kraus, M Larson, and RA Wise. Depth of closure in beach-fill design. In *Proc 12th National Conference on Beach Preservation*, Tallahassee, FL, pp 271-286 (1999).
14. RJ Hallermeier, A profile zonation for seasonal sand beaches from wave climate. *Coastal Engineering.* **4**, 253-277 (1981).
15. RJ Nicholls, M Larson, M Capobianco, and WA Birkemeier. Depth of closure: improving understanding and prediction. In *Proc 26th International Conference Coastal Engineering* (Copenhagen), ASCE, New York, NY, pp 2888-2901 (1998).
16. AJ Bowen, and DL Inman. Budget of littoral sand in the vicinity of Point Arguello, California. Technical Memorandum No 19, USACE Coastal Engineering Research Center (CERC), Ft Belvoir, VA (1966).
17. HL Kaczkowski, and TW Kana. Final design of the Nags Head beach nourishment project using longshore and cross-shore numerical models. In *Proc 33rd International Conference on Coastal Engineering (ICCE)*, 15 pp, Santander, Spain (July 2012).
18. RA Holman, and AJ Bowen, Bars, bumps and holes: models for the generation of complex beach topography, *Journal of Geophysical Research.* **87**, 457-468 (1982).
19. RA Holman, and J Stanley, The history and technical capabilities of ARGUS, *Journal of Coastal Engineering.* **54**(6-7), 477-491 (2007).
20. JC McNinch, Bar and swash imaging radar (BASIR): a mobile x-band radar designed for mapping nearshore sand bars and swash-defined shorelines over large distances, *Journal of Geophysical Research.* **23**, 59-74 (2007).
21. TW Kana, Signatures of coastal change at mesoscales. Coastal Dynamics '95, Proceedings of the International Conference on Coastal Research in Terms of Large Scale Experiments, ASCE, New York, NY, pp 987-997 (1995).
22. JL Davies, *Geographical Variation in Coastal Development.* Hafner Publishing Company (1973).

23 MO Hayes, MO. Barrier island morphology as a function of tidal and wave regime. In S Leatherman (ed), *Barrier Islands*, Academic Press, New York, pp 1-26 (1979).
24. MO Hayes. Georgia Bight. Chapter 7 in RA Davis, Jr (ed), *Geology of the Holocene Barrier Island System*, Springer-Verlag, Berlin, pp 233-304 (1994).
25. R Silvester, and JRC Hsu, *Coastal Stabilization Innovative Concepts*. Prentice Hall (1993).
26. USACE, *Coastal Engineering Manual: Coastal Engineering and Design: Coastal Geology*. USACE, EM 1110-2-1810, Washington, DC (1995).
27. D Reeve, A Chadwick, and C Fleming, *Coastal Engineering: Processes, Theory, and Design Practices.* Spon Press, London (2004).
28 WN Bascom, The relationship between sand size and beach face slope. *Transactions of the American Geophysical Union*. **32**, 866-874 (1951).
29. MO Hayes. Hurricanes as geological agents: case studies of hurricanes *Carla*, 1961, and *Cindy*, 1963. Report of Investigations No 61, Bureau of Economic Geology, University of Texas, Austin (1967).
30. TW Kana. The profile volume approach to beach nourishment. In DK Stauble and NC Kraus (eds), *Beach Nourishment Engineering and Management Considerations*, ASCE, New York, NY, pp 176-190 (1993)
31. JD Rosati, MB Gravens, and WG Smith. Regional sediment budget for Fire Island to Montauk Point, New York, USA. In *Proc. Coastal Sediments '99*, ASCE, New York, NY, pp 802-817 (1999).
32. CSE. Shoreline erosion assessment and plan for beach restoration, Bridgehampton–Water Mill Beach, New York. Feasibility Report for Bridgehampton–Water Mill Erosion Control District, Town of Southhampton, New York. Coastal Science & Engineering, Columbia (SC) (2012).
33. TW Kana, and RK Mohan. 1996. Profile volumes as a measure of erosion vulnerability. In *Proc 25th International Conference Coastal Engineering '96*, ASCE, Vol 3, pp 2732-2745 (1996).
34. CSE. Survey report 2007, Bogue Banks, North Carolina. Monitoring Report for Carteret County Shore Protection Office, Emerald Isle, North Carolina; CSE, Columbia (SC) (2007).
35. TW Kana, HL Kaczkowski, and PA McKee. Myrtle Beach (2001–2010) – Another decade of beach monitoring surveys after the 1997 federal shore-protection project. In *Proc Coastal Engineering Practice 2011*, ASCE, New York, NY, pp 753-765 (2011).
36. CSE. Beach restoration project (2008), Isle of Palms, South Carolina. Interim Monitoring Report – Year 2 (March 2011), City of Isle of Palms, SC. CSE, Columbia (SC) (2011).
37. CSE. Annual beach and inshore surveys—2006 east end erosion and beach restoration project, Kiawah Island, Charleston County, SC. Monitoring Report #5 for Town of Kiawah Island, SC. CSE, Columbia (SC) (2012).
38. CSE. Annual beach and inshore surveys – assessment of beach and groin conditions – survey report 5 — 2006 beach restoration project, Edisto Beach, Colleton County, South Carolina. Report for Town of Edisto Beach, Edisto Island, SC. CSE, Columbia, SC (2011).
39. SB Traynum, TW Kana, and DR Simms, Construction and performance of six template groins at Hunting Island, South Carolina, *Shore & Beach*. **78**(3), 21-32 (2010).

40. M Al-Sarawi, *et al*. Assessment of coastal dynamics and water quality changes associated with the Kuwait waterfront project. Final Report for Environment Protection Council; Faculty of Science, Kuwait University with assistance by CSE, Columbia (SC) (1987).
41. CSE. Beach restoration plan: Jalousie Plantation Resort, St. Lucia. Resort Services Management, St. Lucia; CSE, Columbia, SC (1995).
42. CSE. Shoreline analysis and alternatives for beach nourishment, Rose Hall Beach Hotel, Jamaica. Trammell Crow Hotel Company, Dallas, TX; CSE, Columbia, SC (1990).
43. FEMA. Assessment of current procedures used for the identification of coastal high hazard areas. Office of Risk Assessment, Federal Emergency Management Agency, Washington, DC (1986).
44. TW Kana, and CJ Andrassy. Beach profile spacing: practical guidance for monitoring nourishment projects. In *Proc. 24th International Conference Coastal Engineering*, ASCE, New York, NY, pp 2100-2114 (1995).
45. JD Rosati, Concepts in sediment budgets, *Journal of Coastal Research*. **21**(2), 307-322 (2005).
46. FEMA, *Coastal Construction Manual* – Principals and practices of planning, siting, designing, constructing, and maintaining buildings in coastal areas. Third Edition, US Department of Homeland Security, FEMA 55, Washington, DC (2005).
47. M Crowell, SP Leatherman, and MK Buckley, Historical shoreline change: error analysis and mapping accuracy, *Journal of Coastal Research*. **7**(3), 839-852 (1991).
48. P Bruun, Sea-level rise as a cause of shore erosion, *Journal of Waterways and Harbor Div*, ASCE, New York, NY, **88**(WW1), 117-132 (1962).
49. EB Hands. Predicting adjustments in shore and offshore sand profiles on the Great Lakes. CETA No. 81-4, USACE-CERC, Fort Belvoir, VA (1981).
50. H Hanson, and NC Kraus. GENESIS, generalized model for simulating shoreline change. Tech Rept CERC 89-19, USACE-CERC, Vicksburg, MS (1989).
51. M Larson, and NC Kraus. S BEACH: numerical model for simulating storm-induced beach change. Tech Rept CERC 89-9, USACE-CERC, Vicksburg, MS (1989).
52. NCDENR. Long-term average annual shoreline change study and setback factors. NC Department of Environment and Natural Resources, Raleigh (updated Feb 2004) [see http://dcm2.enr.state.nc.us/maps/ER_1998/SB_Factor.htm] (1998, 2004)
53. CSE. Survey report 2000, Bogue Banks, North Carolina, for Carteret County, Beaufort, NC (2000).
54. CSE. Bogue Banks beach nourishment project – Phases 2 & 3 – Emerald Isle, North Carolina. Preliminary Engineering Report for Town of Emerald Isle, NC. CSE, Columbia (SC) (2002).
55. HJ Verhagen. Method for artificial beach nourishment. In *Proc 23rd International Coastal Engineering Conference*, ASCE, New York, NY, pp 2474-2485 (1992).
56. Moffatt & Nichol. Bogue Banks beach & nearshore mapping program – periodic survey evaluation. Final report, Raleigh, NC (2013) (downloaded 12/2013 from www.protectthebeach.com/monitoring).

CHAPTER 5

TIDAL POWER EXPLOITATION IN KOREA

Byung Ho Choi

Department of Civil and Environmental Engineering, Sungkyunkwan University, 300 Chencheon-dong, Jangan-gu, Suwon, Kyeonggi-do, Korea 400-746
bhchoi.skku@gmail.com

Kyeong Ok Kim

Korea Institute of Ocean Science & Technology, Ansan, Korea 426-744
kokim@kiost.ac

Jae Cheon Choi

Daewoo Engineering and Construction Co., Ltd, Seoul, Korea 110-713
jaecheon.choi@daewooenc.com

The highest tides in South Korea are found along the northwest coast between latitudes 36-38 degrees and the number of possible sites for tidal range power barrages to create tidal basins is great due to irregular coastlines with numerous bays. At present Lake Sihwa tidal power plant is completed. The plant is consisted of 10 bulb type turbines with 8 sluice gates. The installed capacity of turbines and generators is 254MW and annual energy output expected is about 552.7 GWh taking flood flow generation scheme. Three other TPP projects are being progressed at Garolim Bay (20 turbines with 25.4MW capacity), Kangwha (28 turbines with 25.4MW capacity), Incheon (44 or 48 turbines with 30 MW capacity) and project features will be outlined here. The introduction of tidal barrages into four major TPP projects along the Kyeonggi bay will render wide range of potential impacts. Preliminary attempts were performed to quantify these impacts using 2 D hydrodynamic model demonstrating the changes in tidal amplitude and phase under mean tidal condition, associated changes in residual circulation (indicator for SPM and pollutant dispersion), bottom stress (indicator for bedload movement), and tidal front (positional indicator for bio-productivity) in both shelf scale and local context. Tidal regime modeling system for ocean tides in the seas bordering the Korean Peninsula is designed to cover an area that is broad in scope and size, yet provide a high degree of resolution in strong tidal current region including off southwestern tip of the Peninsula (Uldolmok , Jangjuk, Wando-Hoenggan), Daebang Sudo (Channel) and Kyeonggi Bay. With this simulation system, real tidal time simulation of extended spring-neap cycles was performed to estimate spatial distribution of tidal current power potentials in terms of power density, energy density and then extrapolated annual energy density.

1. Introduction

The highest tides in South Korea are found along the Northwest coast between latitudes 36 and 38 North. The coastline in this region is very irregular with numerous bays and inlets of varying sizes. The number of possible sites for tidal power barrages to create reservoirs is therefore great. In addition, a number of islands off the coast offer the potential for use as bases in the construction of major tidal power barrages enclosing very large reservoirs.

2. Korea Tidal Power Study in 1978

Some ambitious tidal power schemes have been proposed at "oceanic" sites within Kyounggi Bay. For such schemes, parts of the Bay would be enclosed by constructing dikes between several offshore islands. Because of the sensitivity of a tidal system to alterations, such major schemes can at present not be realistically considered, without detailed and reliable model studies. Such schemes would very likely destroy to a large extent the tidal movements which they would be intended to exploit.

On considering the maximum feasible tidal power potential on Korea's west coast, it has to be recognized that the existing high tides are caused by a number of complex phenomena. By building tidal power barrages the amplitude of the tides could change. Choi (1981) performed modeling of changes in tidal regime by computer techniques, so called as tidal regime modeling. Figure 1 illustrates the maximum feasible tidal power potential (Table 1) which can be realistically identified at the time of investigation. The concentrated study areas are listed on Table 1. Seogmo-Do, borders to the north on the Han River which is at this point the southern boundary of the DMZ separating South and North Korea. Other study areas to the south are arranged along Kyonggi Bay and continue along the shore line of the Taean Peninsula where various bays and inlets provide opportunities for the development of tidal power. Cheonsu Bay represents the most southerly tidal power scheme considered.

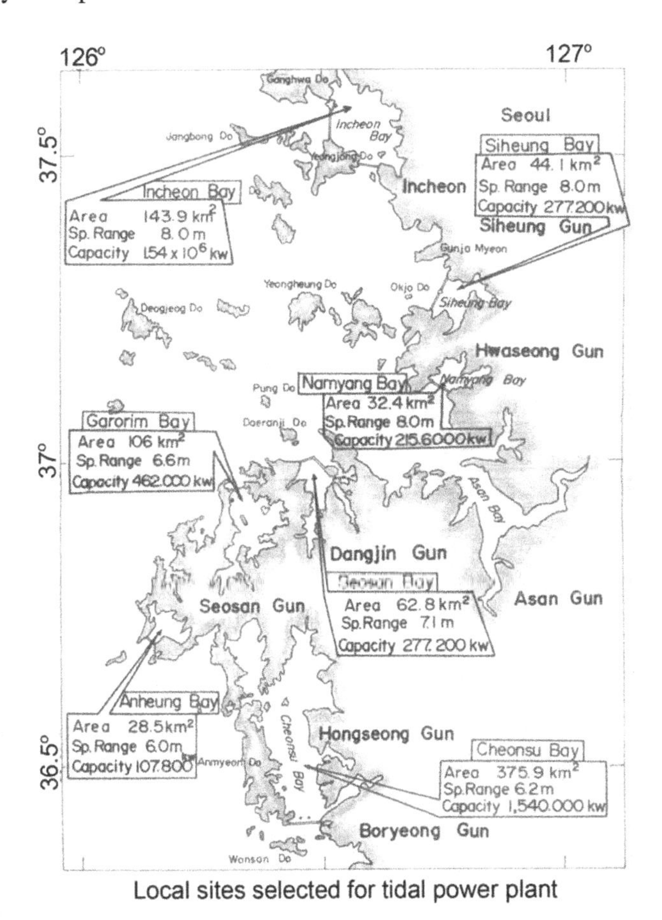

Fig. 1. Location of tidal range power plants investigated in the past (KEC, 1978).

Table 1. Single lined table captions are centered to the table.

Location	Output (10^3kW)	Generation (10^6kWh)	Ave. (Max.) tidal range (m)	Dike length (km)	Reservoir area (km^2)
Ocean tide					
Seokmo Island	1,140	2,892	5.4 (8.8)	18.3	216.9
Sin Island (Out)	810	2,079	5.6 (9.6)	23.0	192.3
Sin Island (In)	(660)	(1,650)	5.6 (9.6)	17.6	120.0
Yeongheung Island	1,800	5,102	5.5 (8.5)	19.9	657.5
Sub total	3,750	10,073			
Coastal tide					
Incheon Bay	330	900	5.7 (9.2)	5.1	79.6
Asan Bay (Out)	810	2,229	6.1 (9.6)	6.6	151.7
Asan Bay (In)	(450)	(1,345)	6.1 (9.6)	2.4	103.4
Seosan Bay	180	412	5.4 (7.7)	3.2	56.2
Garolim Bay	330	820	4.8 (7.9)	2.1	120.0
Cheonsu Bay	540	1,239	4.5 (7.6)	5.1	350.5
Sub total	2,190	5,600			
Total	5,940	15,673			

3. Tidal Range Power Development Plan in 2010

Four tidal range power development plans are underway in South Korea. The main purpose of this project is to provide an alternative, clean energy source (tidal energy) using a conventional tidal power generation method (Bernstein, 1965; Bernstein et al.,1997) where an estuary is barraged with a dam, a technique first suggested by Struben (1921). The first Korean tidal power plant that went into operation is the Lake Sihwa Tidal Power Plant. The Lake Sihwa tidal barrage makes use of a seawall constructed in 1994 for flood mitigation and agricultural purposes. Ten 25.4 MW submerged bulb turbines are driven in an unpumped flood generation scheme; power is generated on tidal inflows only and the outflow is sluiced away. However, it is well known that flood generation schemes, which are usually adopted to secure land within the lake for development, have the disadvantage that the useable volume of water between the low tide and mid tide in the lake is much less than that created by an ebb generation scheme that utilizes the volume of water available between mid tide and high tide. Furthermore, a large portion of the inter tidal areas are permanently exposed with flood generation schemes, resulting in a rapid change from marine ecosystem to a fresh water ecosystem, contrary to the claims that a marine ecosystem can be preserved (Baker, 1991). Meanwhile, Prandle (1984) has shown that a two-way mode can be operated at a more constant head where the flushing regime in the enclosed tidal basin remains closer to the undisturbed state, but this aspect cannot be exploited due to practical engineering realities. The other three projects are under construction, which are Garorim, Incheon and Ganghwa Tidal Power Project. Figure 2 and Table 2 show the location and detail information of each tidal range power plant.

4. Sihwa Lake Tidal Power Plant

The construction of a Tidal Power Plant (TPP) required temporary circular cell cofferdams to enable dry work. The TPP construction on Lake Sihwa is similar to the conventional La Rance project except that there is no precast concrete caisson structure. The Lake Sihwa cofferdams were built without supporting rangers, and their stability is provided solely by the cell filling. While the standard length of a circular cell cofferdam is 28 m, the required length is up to 31.5 m for the Sihwa project due to the water depth and ground conditions. The circular cell cofferdam at Sihwa consists of 29 primary cells and 28 spandrel walls; welded distribution piles connect the primary cells and spandrel walls. The construction of a circular cell was performed offshore with the aid of a control desk. The flat sections were lifted by crane from a pontoon and driven to the required depth at the site with the aid of vibrating hammer.

First, the silt protection, cellular cofferdam and diversion roads were constructed. Then dewatering and excavation were performed for dry work, and the main concrete structures for power plants and sluices were constructed. Then the turbo-generators and sluice gates were installed after unloading from a temporary wharf to the erection bay. These machines were then transported by a special carrier to a gantry crane and a hydraulic crane. This construction schedule was necessary considering the non-uniform settlement of the railway foundation and to ensure that the gantry crane installed in July 2009. After a new road is constructed over the sluice gates, the diversion roads built and detached. Then the cellular cofferdam was removed, and the turbo-generators were tested for normal TPP operation.

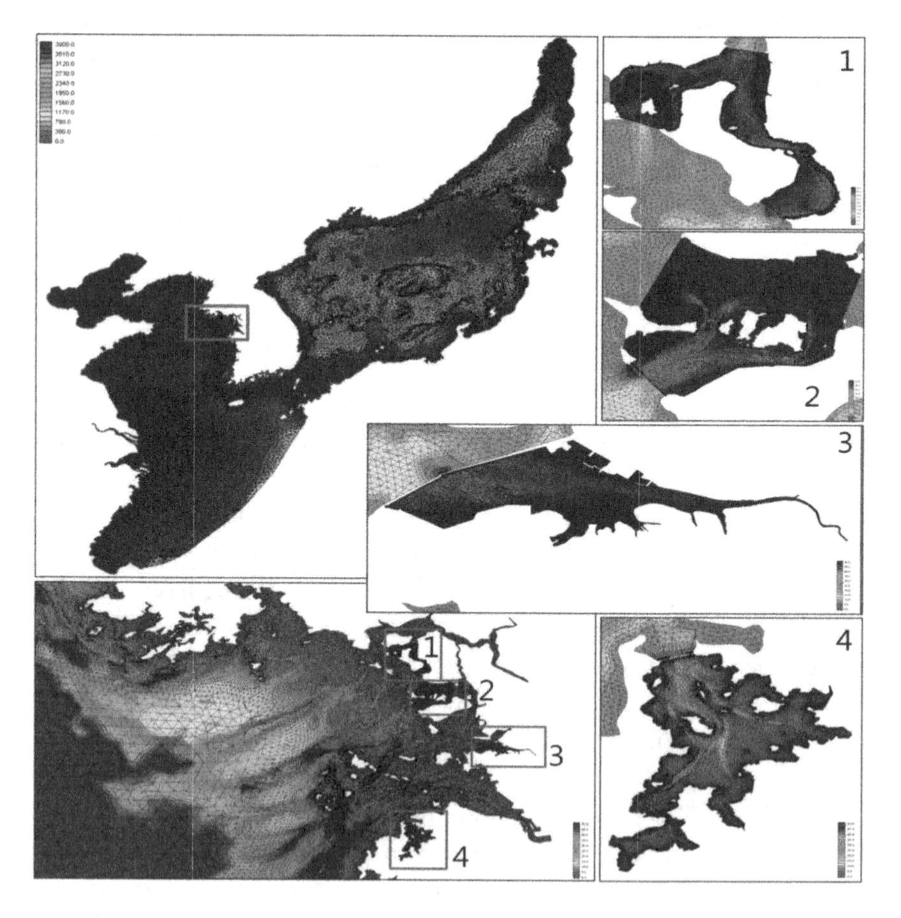

Fig. 2. Mesh system for simulation of tidal range power plant constructions in Kyeonggi Bay.

Table 2. Tidal range powers in the recent investigation.

Location	Gangwha	Incheon	Sihwa	Garolim
Generation type	One-way ebb	One-way ebb	One-way flood	One-way ebb
Location outline				
Max. tidal range (m)	8.97	9.095	7.80	8.14
Reservoir area (km^2)	85	157.5	42.4	96
Reservoir volume ($10^6 m^3$)	730	1,520	324	650
Energy output & dike outline				
Capacity (MW)	813	1,320	254	520
Annual generation (GWh)	1,536	2,414	552.7	950
Dike length (m)	6,500	18,300	12,676	2,053

N.B. Except Lake Sihwa Plant completed, ther three projects are delayed as of 2013.

Fig. 3. Photograph showing the construction progress of Lake Sihwa Tidal Power Plant.

Fig. 4. Machinery setup in 2010 showing 10 turbo-generators and eight sluice gates.

The construction of the circular cell cofferdam completed, immediately after dewatering and excavation in October 2007, concrete structures have also been constructed to house the turbines and gates inside the cofferdam. Figure 3 shows the photographs of the construction through December 2008. Figure 4 shows the machinery setup in 2010. Real time tide prediction for the localized TPP construction site is able to predict the electricity that can be generated and the tidal level at the outside and inside of the dike wall.

The Sihwa tidal power plant is designed as a flood generating system, taking advantage of the difference in the tide levels between the sea and the artificial lake. Flood generating systems create power from the incoming tide, i.e., the water flowing from the sea to a basin. When the high tide enters, water flows through the turbines to create electricity. Separate gates beside the turbines are designed to open during the ebb phase. When the low tide enters, the gates are raised, and the water flows out. The turbines operate in a sluicing mode during the ebb phase, and no energy is produced.

The characteristics and methods of TPP computer simulations are shown in Fig. 5. An analysis of the optimum energy modes of a TPP should include the actual conditions under which the TPP operates, such as parameters defining its hydraulic and power equipment, hydraulic engineering structures, basin and power system. Such considerations necessitate a large number of data to be inputted into the computer. Some of these parameters can be obtained by experimental measurements, which by nature contain some errors.

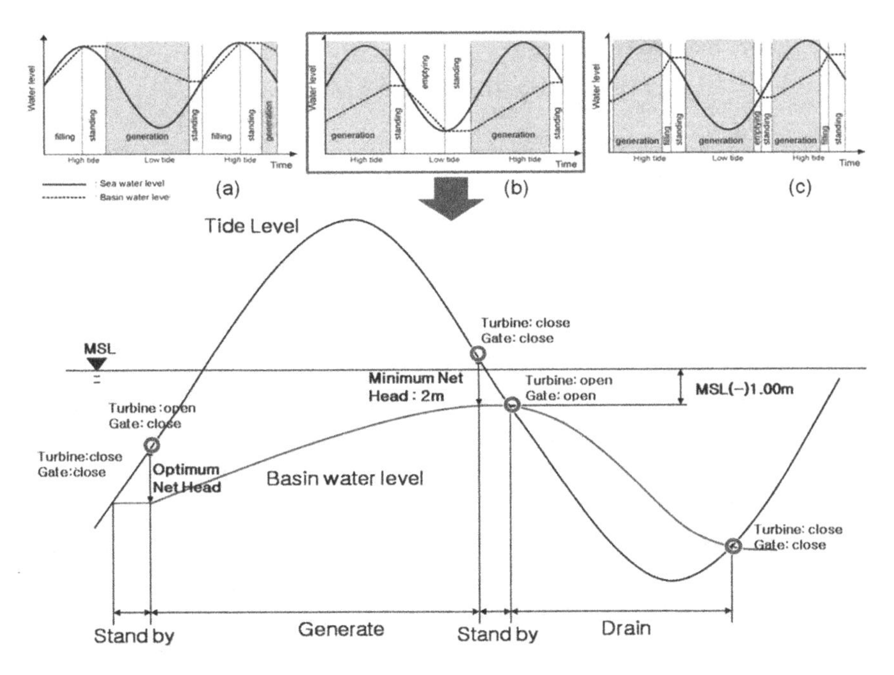

Fig. 5. Three type of tidal power generation; (a) single effect operation (ebb), (b) single effect operation (flood), and (c) double effect operation. And detail schematic diagram of "Single effect operation (flood)" selected in Sihwa Tidal Power Plant.

The values of the function for arguments other than those specified in the table are computed by linear interpolation. Usually it is advisable to use this method for parameters that are the function of one variable because when a function in the model contains two variables, the data table becomes too large, particularly when high accuracy is required. Furthermore, the computational time increases for parameters with multiple variables. For parameters with multiple variables, the values can be pre-approximated with a polynomial raised to the power m in the form:

Electric Generation during flood:

$$Q = 0 \qquad (0 < \Delta H < 1.07)$$

$$Q = 150.433 \times \Delta H + 113.248179 \qquad (1.07 \le \Delta H < 1.95)$$

$$Q = 18.1077 \times \Delta H + 371.3606 \qquad (1.95 \le \Delta H < 5.69)$$

$$Q = 1767.8881 - 346.5186 \times \Delta H + 20.9571 \times \Delta H^2 \qquad (5.69 \le \Delta H < 7.43)$$

Power generated (MW):

$$P = -1.7256 + 0.3161 \times H + 1.2465 \times H^2 - 0.0907 \times H^3 \qquad (\Delta H < 5.64)$$

$$P = 22.29 \qquad (\Delta H \geq 5.64)$$

Discharge (through Sluice gate):

$$Q = 800 \times \Delta H \qquad (0 < \Delta H < 0.05)$$

$$Q = 600 \times \Delta H + 10 \qquad (0.05 \leq \Delta H < 0.1)$$

$$Q = 250 \times \Delta H + 45 \qquad (0.1 \leq \Delta H < 0.3)$$

$$Q = 59.56939963 + 231.1052218 \times \Delta H - 94.5018523 \times \Delta H^2 + 26.29460961 \times \Delta H^3 - 2.754795024 \times \Delta H^4 \qquad (0.3 \leq \Delta H < 2.5)$$

$$Q = 839 \times \Delta H^{0.3855} \qquad (\Delta H \geq 2.5)$$

The approximation of the parameter and the polynomial degree required are both determined with a computer program. This program has a criterion for a minimum R.M.S. deviation for the approximating polynomial based on the values of the original parameter at the points of approximation.

The following section describes the parameters used to calculate the optimal TPP operating regime. The discharge characteristic of a TPP generating unit is specified for four modes of operation: FT (Forward Turbine), BT (Backward Turbine), FP (Forward Pumping), and BP (Backward Pumping). This characteristic represents the power rating, Nu, of the TPP generating unit as a function of the discharge Q through the unit and the head H. The discharge characteristic is approximated by a polynomial. The unit of generated power, P, by turbine during flood is Megawatt. The reversible double-effect generating unit design allows operation in the FT mode to be easily changed to operation in the BP mode (and from BT to FP) without altering the direction of rotation. Therefore, only two polynomials, one for the modes FT and BP and another for the modes BT and FP, are required to approximate the discharge characteristic. In order to make multiple variable TPP computations feasible, the mathematical model is used to provide computer-aided recalculations from the equations. These recalculations are based on the law of proportionality, and they provide the discharge characteristics of the TPP generating unit specified in the source data in terms of the diameter and rotational speed.

Figure 6 shows the water elevation in sea side and lake side, and difference of both during the pre-operation period. Figure 7 shows the generated power in unit of MW at the first generator during July 2012, still in pre-operational stage. The difference of the water elevation in sea side and lake side is shown together for comparing. Currently the plant is completed, but the complete operation records are not available because still pilot services for evaluation.

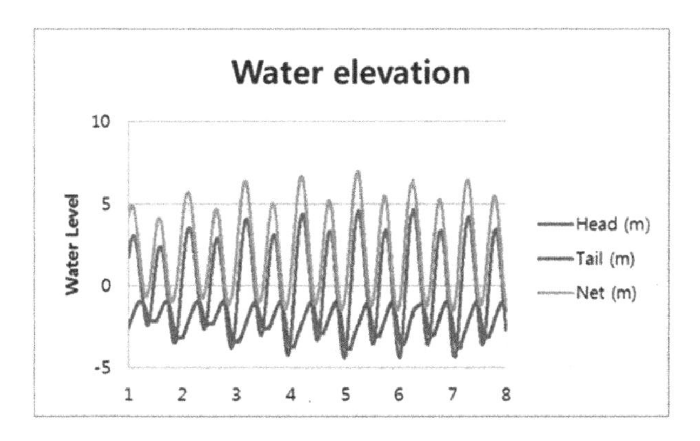

Fig. 6. Water elevation in sea side (head) and lake side (tail) during July 2012.

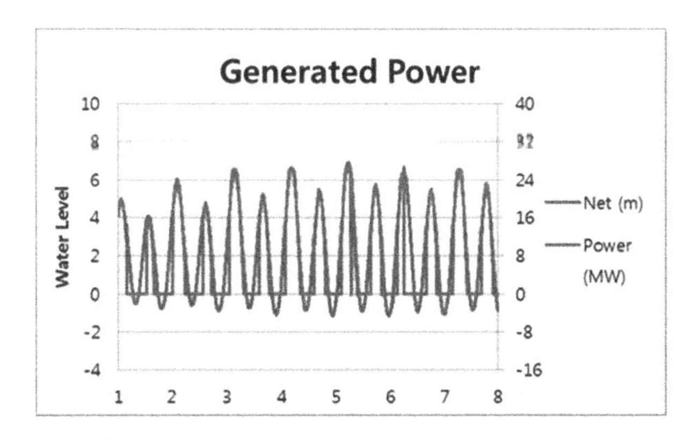

Fig. 7. Difference of water elevation and generated power at the first generator during July 2012.

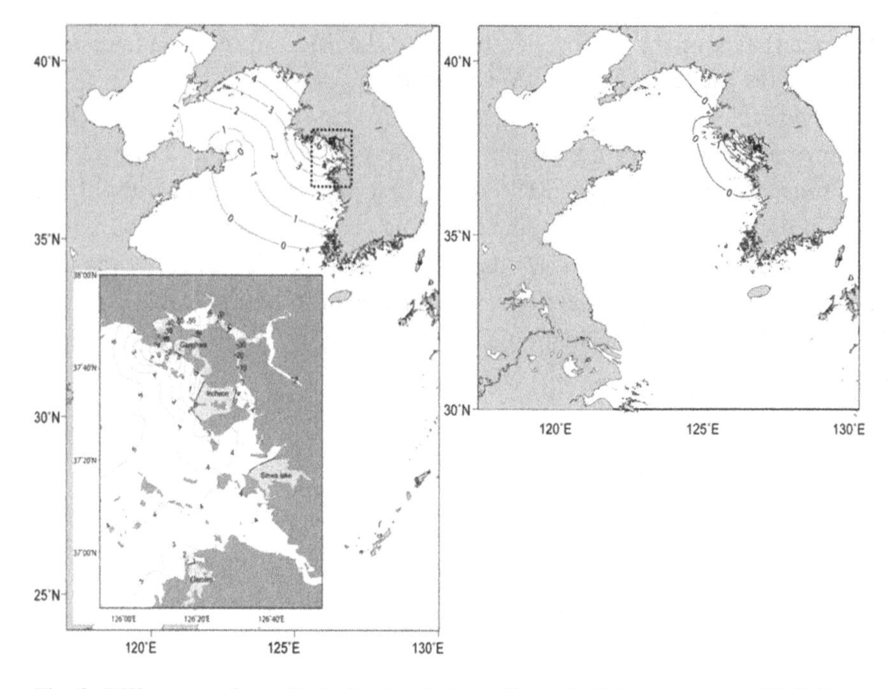

Fig. 8. Differences of amplitude (cm) and phase (degree) tidal component of M2 due to barrage construction.

5. Hydro-Environmental Impacts of Power Schemes in Kyeonggi Bay

The unstructured fine mesh system was designed using a 20 meter minimum grid size along the four disks at Sihwa, Incheon, Gangwha and Garolim. The 1 arc-second land and sea combined elevation and bathymetry dataset (SKKU1sec, revised positional system with WGS84) allows time-efficient preparation of basic inputs. Here we examine the specific potential impacts of tidal power schemes in the near and far-fields for Kyeonggi bay. ADCIRC model allows the simulation of multiple barrage schemes in conjunction to explore their cumulative impact upon the environment and their power production potential. The unstructured grid permits the resolution to change from a coarse 5 km in the deep water at the forced open boundary down to 20m in shallow estuaries and near barrage structures. The grid contains almost 460,000 nodes and over 890,000 elements are solved with a 1second time step. This takes 5 days for 24 day simulation which includes 10 day spin up and a 14 day settling period and was forced using the dominant tidal constituent (M2) in this region. The model was first run without barrages, for a baseline state, and then with barrages of four tidal power plants within the Kyeonggi Bay.

Study has been used to examine large-scale changes in tidal amplitude and bed stress for sediment transport (without recourse to a detailed sediment transport model) and other environmental changes. The difference of simulated amplitude after the construction dikes as provided in Fig. 8 showing that perturbations are occurred over the limited regional system, increased amplitudes about from 3 cm to 6 cm in M2 tide in the Kyeonggi Bay. In case of phase disturbance it is found that the difference about 5 degree in M2 tide is shown. In particular the difference of phase about 4 degrees in the M2 tide after the construction of dikes occur at Sihwa compared to previous situations without the dikes.

The bed stress is an indicator of sediment movement as bed load and various regions of convergence and divergence are often in good agreement with locations of sediment deposition (Wolf et al., 2009). The bottom shear stress (τ_b) is a combined effect of currents and waves being induced by the bottom friction which is influenced by bottom roughness. The bottom shear stress which is an index for us to make it possible to evaluate changes and deformation of tidelands was estimated as follows;

$$\tau_b = \rho_w f_b u|u|,$$

where ρ_w : density of water (kg/m3), f_b : friction coefficient, and u : depth-averaged velocity (m/s). Coefficient of the bottom friction (f_b) is given as follows (Dyer, 1986);

$$f_b = 2\frac{\mathrm{k}^2}{\left(\ln\left(30H/k_N - 1\right)\right)},$$

where k : Von Karmann constant (~0.4), H : water depth (m), and k_N : bottom roughness (m).

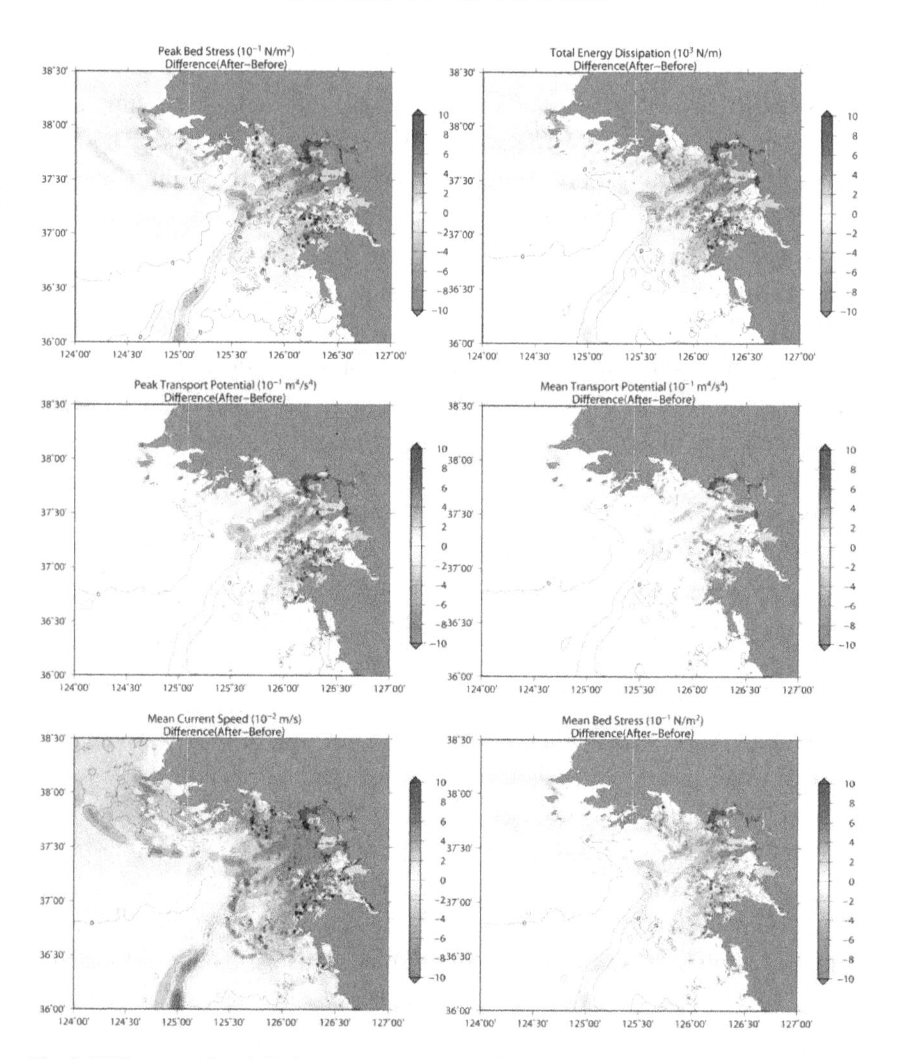

Fig. 9. Difference of peak bed stress, total energy dissipation, peak transport potential, mean transport potential, mean current speed and mean bed stress due to barrage construction.

The bedload sediment transport rate is expected to be related to some power of the bed stress. The difference of M2 bottom stress distributions before and after the dikes is shown in Fig. 9, which reveals the location of the major areas of energy dissipation. The reduction in bottom stress seen in the Kyeonggi Bay has implications for the dynamic sediment regime in this area. The reduced bottom stress will become less turbid and allow more light penetration. Two more parameters such as tidal energy dissipation and sediment transport potential were also investigated and presented to help the understanding of the effects of barrier construction and energy environment. The tidal energy dissipation was estimated from the application results of numerical simulations by converting the maximum bottom shear stress by bottom friction into work per unit area. It is known that the sediment transport potential which represents the sediment transport capacity is in proportion to the third or fourth power of maximum currents velocity (Choi, 2001). Therefore, the sediment transport potential was estimated by the fourth power of maximum currents velocity in this study.

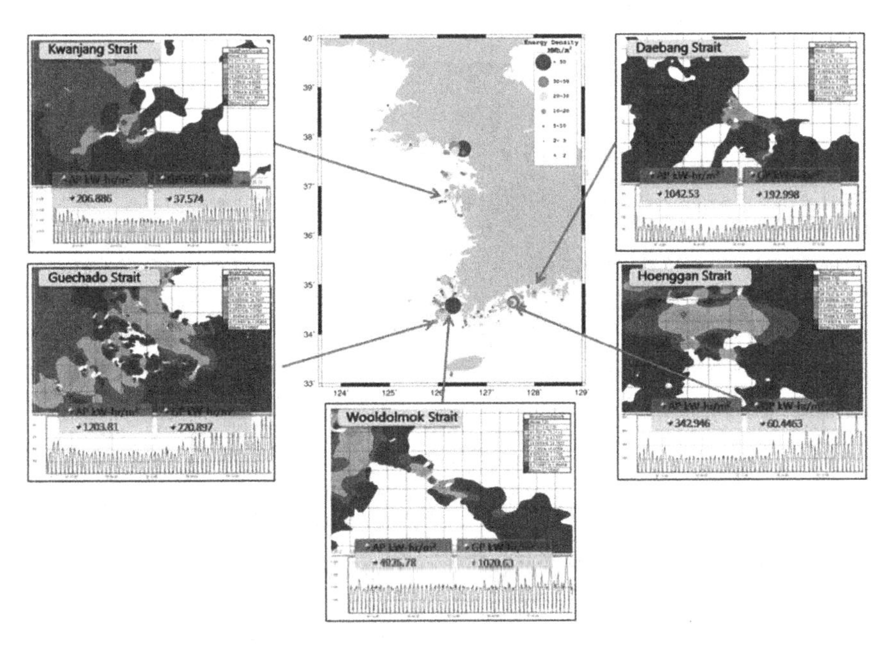

Fig. 10. Estimated Tidal Current Energy Density which was estimated by Blue Kenue (CHC-NRC, 2010: www.nrc-cnrc.gc.ca/eng/ibp/chc/software/kenue/blue-kenue.html).

6. Tidal Current Power Potential

Tidal current power generations have been studied in Korea, also they developed turbine and plants that generate electricity through the movement of tidal currents. The difference between a tidal plant and a tidal current power plant is that the latter requires no dams and can be operated almost 24 hours a day. The tidal power density and energy density can be estimated by numerical simulation. Figure 10 shows the estimated tidal current energy density which was estimated by Blue Kenue (CHC-NRC, 2010). The speed of tidal current at the Uldolmok (Myeong-Ryang) strait is among the fastest in the Korea. This area could effectively produce an additional 4900 kW-hr/m3. Figure 11 shows the location of Uldolmok Pilot TCPP and installed turbines in in-situ tests. The construction of pilot plant of 1MW has been started in April 2006, and is completed in 2007 (Yum, 2007).

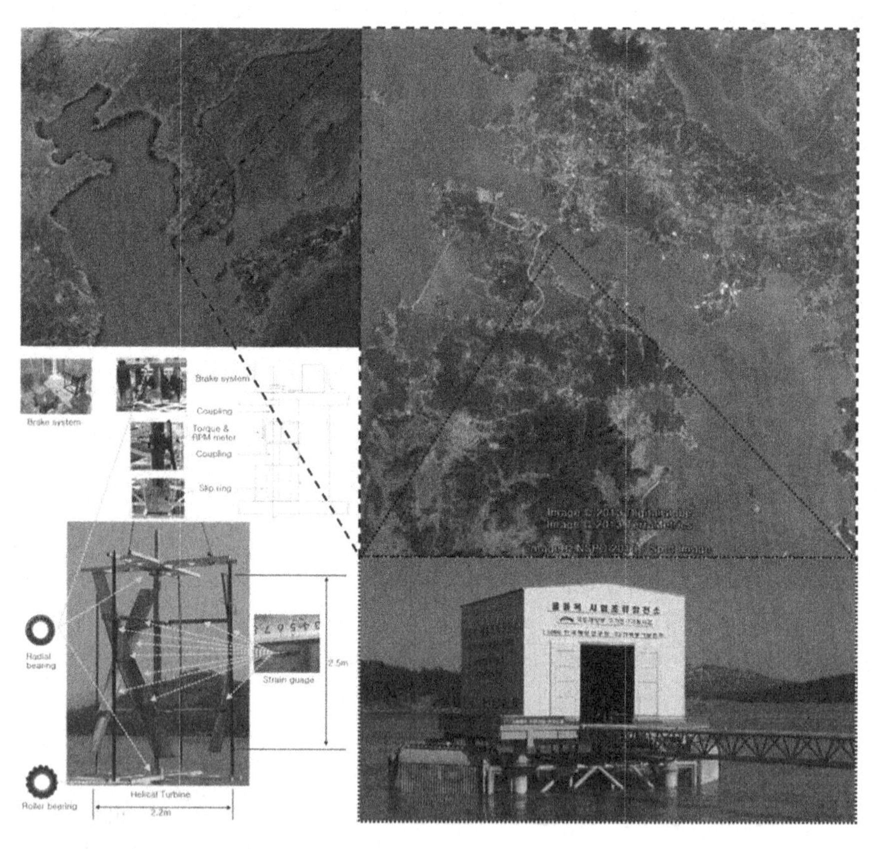

Fig. 11. Location of Uldolmok (Myeong-Ryang) Pilot TCPP, tested Helical Turbine for 1 MW pilot plant.

7. Conclusion

This paper briefly outlined status of tidal exploitation in Korea. After successful construction of Lake Sihwa Tidal Power Plant, subsequently there is a possibility of building Garolim Tidal Power Plant. Tidal current power plant as 1 MW pilot station (Uldolmok) is not extending to further substantial development.

The introduction of tidal barrages into 4 tidal power projects along the west coast of the Korea Peninsula may impose awide range of potential impacts. Here we have attempted to quantify these impacts using 2-D models. The largest potential impact at the far field scale is the increase in tidal amplitude possibly leading to enhanced coastal flood risk. There is an associated change in bottom stress with each barrage which will impact upon the benthic habitats at each site. Tidal current energy density has been estimated by 2-D numerical models and several areas could effectively produce additional electricity energy if current power turbine-generators are proven to be feasible.

References

Baker, A. C. (1991). *Tidal Power.* Institution of Electrical Engineers, Peter Peregrinus Ltd, London, 250p.

Bernstein, L. B. (1965). *Tidal Energy for Electric Power Plants.* Israel Program for Scientific Translations, Jerusalem, 344p.

Bernstein, L. B., Wilson, E. M. and Song, W. O. (1997). *Tidal Power Plants.* Korea Ocean Research and Development Institute, Ansan, Korea.

Choi, B. H. (1980). *Analysis of Tide at Incheon.* Korean Ocean Res. Development Inst. Report 80-01.

Choi, B. H. (1981). *Effect on the M2 Tide of Tidal Barriers in the West Coast of Korea.* Korean Ocean Res. Development Inst. Report 81-01.

Choi, B. H. (2001). Effect of Saemangeum tidal barriers on the Yellow Sea tidal regime, *Proceedings of the first Asian and Pacific Coastal Engineering Conference, APACE 2001*, Dalian, China, 1, pp. 25-36.

Chosun Governer's Office (1930). Tidal Power. Bureau of Communication. (in Japanese)

Dyer, K. R. (1986). Coastal and Estuarine Sediment dynamics. John Wiley & Sons. Chichester. 342 p.

Gibrat, R. (1966). L'energie des marees. Presses Universitaires de France, Paris.

Kinnmark, I. P. E. (1985). The Shallow Water Wave Equations: Formulation, Analysis and Application. Lecture Notes in Engineering Vol. 15. Springer-Verlag, New York, 187p.

Korea Electric Company (1978). Korea Tidal Power Study - Phase I. KORDI, KIST, Shawinigan Engineering Co., 180p.

Korea Electric Power Corporation (KEPCO) (1993). Feasibility Study on Garolim Tidal Power Development. Korea Ocean Research and Development Institute, Ansan, Korea.

Luettich, R. A., Westerink, J. J. and Scheffner, N. W. (1992). ADCIRC: An Advanced Three-dimensional Circulation Model for Shelves, Coasts, and Estuaries, Report 1: Theory and Methodology of ADCIRC-2DDI and ADCIRC-3DL. Dredging Research Program Technical Report DRP-92-6, U.S. Army Engineers, Waterways Experiments Stations, Vicksburg, MS, 137p.

Lynch, D. R. and Gray, W. G. (1978). A wave equation model for finite element tidal computations. Computers and Fluids, 7, pp. 207-228.

Prandle, D. (1984). Simple Theory for Designing Tidal Power Schemes. Adv. Water Resources, 7, pp. 21-27.

Struben, A. M. A. (1921). Tidal Power. Sir Isaac Pitman & Sons, Ltd., London.

NRC-CHC (2010). Reference Manual for Blue Kenue, Canadian Hydraulics Centre National Research Council, 207 p.

Wolf, J., Walkington, I. A., Holt, J. and Burrows, R. (2009). Environmenatla impacts of tidal power schemes. Proceedings of the Institution of Civil Engineers-Maritime Engineering, 162 (4), pp. 165-177.

Yum, K. D. (2007). Tide and tidal current energy development in Korea, Proceedings of the 4th International Conference on Asian and Pacific Coasts, APAC2007, Nanjing, China, pp. 42-55.

CHAPTER 6

A FLOATING MOBILE QUAY FOR SUPER CONTAINER SHIPS IN A HUB PORT

Jang-Won Chae and Woo-Sun Park

Coastal Development and Ocean Energy Research Division, Korea Institute of Ocean Science and Technology. Ansan

P.O. Box 29, Seoul 425-600, Korea.
E-mail: jwchae@kiost.ac, wspark@kiost.ac

A floating mobile quay (FMQ), which is an innovative berth system, has functions of not only both side loading/unloading but also direct transshipment to feeder ships in a hub port. Applying the FMQ to a hub port such as the west terminal of Busan New Port of Korea, it is shown from a physical modeling and field model test that the quay is dynamically stable and workable in the prevailing wave condition and also safe in a design storm condition, respectively. The terminal productivity is increased by 30% comparing with the present land based berth. The B/C ratio of the new berth system is evaluated as 1.13 considering super-large container ships. It appears that the FMQ is a technically and economically feasible system in the hub port.

1. Introduction

The billion TEU mark of worldwide terminal throughput would probably be surpassed in 2007 (refer to a new market report by German company OneStone, 2006, World Port Development Nov. 2006). It was 625 million TEU in 2010 and will increase a little slowly to about 885 million TEU in 2015 due to the financial crisis. According to the report the throughput growth is dependent to a large extent on the transshipment rate and empty container handling rate. Highest market growth is in North East Asia (NEA). The transshipment (TS) cargos in

2011 and 2020 are estimated as 1,320,000 TEU and 2,100,000 TEU, respectively, and the rate is larger than 40% for the domestic shipment in Korea. As a hub port the Busan New Port is being developed in order to cope with the increasing container cargo.

Container ships (CS) in 7,000 ~ 8,000 TEU class are being operated in main (trunk) routes recently and a container ship in 15,500 TEU class of Emma Maersk was launched in the Europe-South East Asia route, September 2006. It is expected that very large container ships (VLCS) in 18,000 TEU class will be operated in the main trunk route in 2013. It will be a hot issue how a terminal can serve VLCS in a day (24 hrs), because reducing the ship residence time in a harbor can reduce one ship from its liner's fleet and that is one of important requirements to be a major hub port.

However, current harbor facility and cargo handling (loading/unloading, transportation) system cannot handle T/S of super large container ship within 24 hours, which will be crucial for the ship calling. Therefore, future container harbor should accommodate innovative solutions and flexibility in capacity expansion in sustainable way[1]. This paper deals with a new concept of harbor layout and structures such as both side loading/unloading system and, also the latest high efficient/intelligent container crane (CC) and cargo handling system in order to increase harbor productivity for the requirements of the VLCS calling. A floating mobile quay (FMQ) may be an efficient alternative for loading and unloading containers at both sides in combination with the existing land based fixed quay wall (QW).

Relevant research works have been carried out such as floating crane and quay[2] and floating terminal[3]. Valdez floating concrete terminal (213 m x 30.5 m) has been operated successfully for 24 years now. Other examples are a Mega Float Project for floating airports of Japan[4,5,6] and Mobile Offshore Base for military purposes of US Navy[7,8,9].

A new berth system with a floating quay is developed to increase loading and unloading capability of the existing land based berth, dramatically. The floating quay can supply two major functions to the new berth system. One is a both side loading/unloading function, and the other is a direct transshipment function to feeder ships. In addition, the floating quay is designed as a mobile quay to improve the quay productivity significantly [10].

As an example, a FMQ system is proposed for the west terminal of Busan New Port of Korea, and evaluations were made on the safety, stability, and workability of the quay, and productivity and economic feasibility of the system. For this proposal extensive physical modeling and designing process have been done[11].

2. Floating Mobile Quay (FMQ)

2.1. *Definition*

The FMQ (in other words Hybrid Quay Wall, HQW) is basically consisted of a very large pontoon type floating structure with thrusters controlled by the dynamic positioning technique, mooring systems, and access bridges to an existing land side berth. It has functions of both side loading/unloading and direct transshipment to feeder vessels in

(a) Isometric view

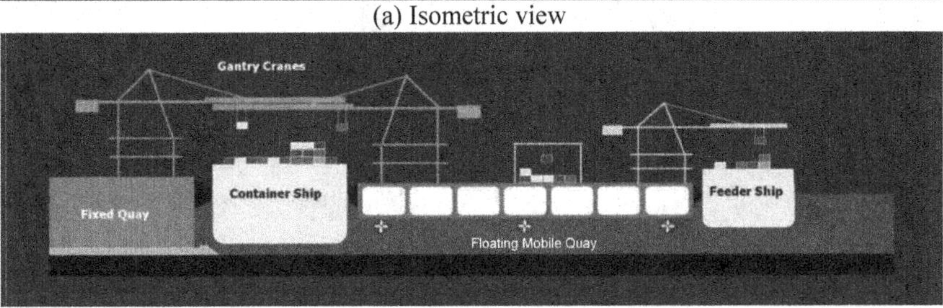

(b) Cross sectional view

Fig. 1. View of a terminal adopting the FMQ.

combination with the existing land based berth, container crane (CC), transfer crane (TC), automated lifting vehicle (ALV) and an intelligent operating system for the FMQ.

Isometric and cross sectional views of the terminal adopting the FMQ are shown in Fig. 1. Two different applications of the FMQ are considered. One is a movable quay along an existing berth and the other is an indented berth with adjustable width. Fig. 2 shows conceptual operation diagrams for two applications.

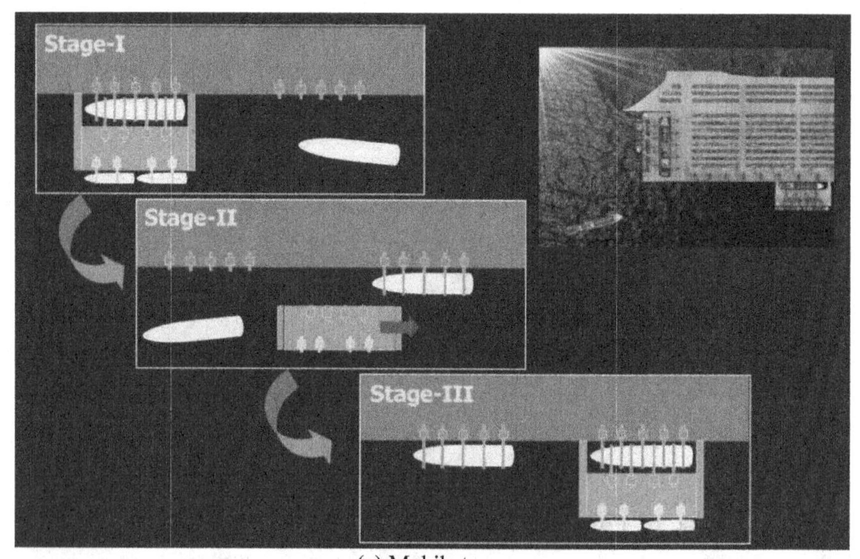

(a) Mobile type

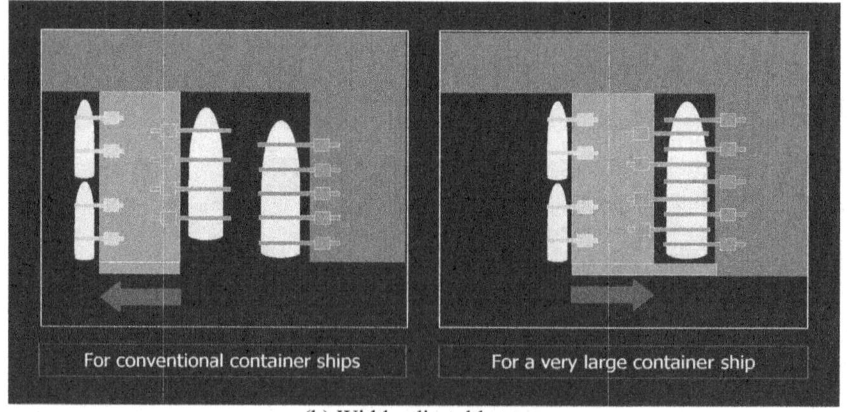

(b) Width adjustable type

Fig. 2. Conceptual operation diagrams for berth systems adopting the FMQ.

The width of the indented berth can be enlarged to more than twice of a usual container ship's (6,000 ~ 8,000 TEU) width in order to accommodate two container ships when the berth is off-duty for a VLCS of 15,000 TEU. The FMQ of the movable berth can travel along an existing aligned 2 ~ 3 berths in order to maximize the productivity of the berth system.

2.2. *Performance Analysis*

Stability and Workability

To investigate the stability and workability of the FMQ for operational and storm conditions, hydraulic model test was carried out in a scale of 1/100. Two container ships were considered with the FMQ. One is CS of 6,750 TEU as mother ship, and the other is CS of 4,500 TEU as feeder ship. Their dimensions are shown in Table 1.

The hydraulic model test has been performed for 3 cases as shown in Fig. 3 by using various regular and irregular waves in Table 2. To measure hydraulic and dynamic responses of the system, 30 instruments were used, i.e., 11 wave gauges, 4 accelerometers on top of container cranes, 4 fender force meters, 8 mooring line tension gauges, and FMQ and two container ship's motion response measuring system RODYN.

Table 1. Dimensions of the FMQ and container ships.

Type	Length	Width	Depth	Draft	K_G
	(m)	(m)	(m)	(m)	(m)
FMQ	350.0	160.0	7.0	5.0	4.77
CS1 (6,750 TEU)	286.6	40.0	15.0	11.98	17.37
CS2 (4,500 TEU)	250.0	36.3	15.0	13.6	13.6

(a) Case 1 (QW+FMQ)

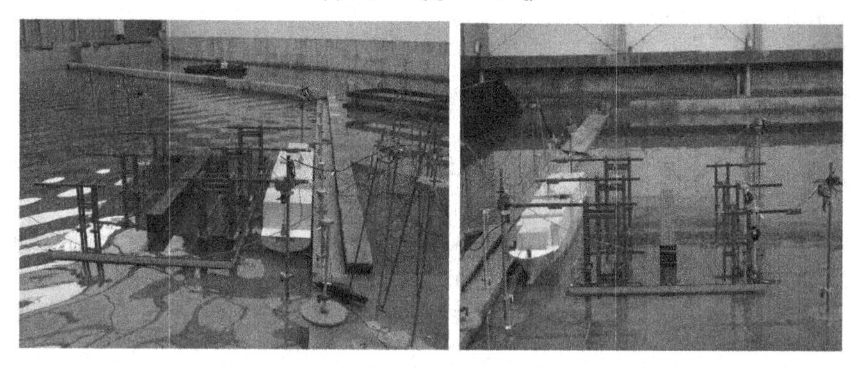

(b) Case 2(QW+CS1+FMQ)

(c) Case 3 (QW+CS1+CS2+FMQ)

Fig. 3. Three cases for hydraulic model test of FMQ applications.

Table 2.Wave conditions for model test.

Wave	Regular	Irregular	
Height (m)	1.0	1.0	1.5
Period (s)	7, 9, 17, 21, 25	7, 12, 15	7, 12, 15
Attack angle (degree)	45, 90	45, 90	

Accelerations of FMQ in response to the regular waves are shown in Fig. 4. The criteria of crane acceleration is used as same as that suggested from Mega-flat study in Japan, which are equivalent as the acceleration limit of land based crane. As for 5 seconds motion it is referred as 0.16 m/s^2 (RMS) and for 10 seconds as 0.32 m/s^2 (RMS).

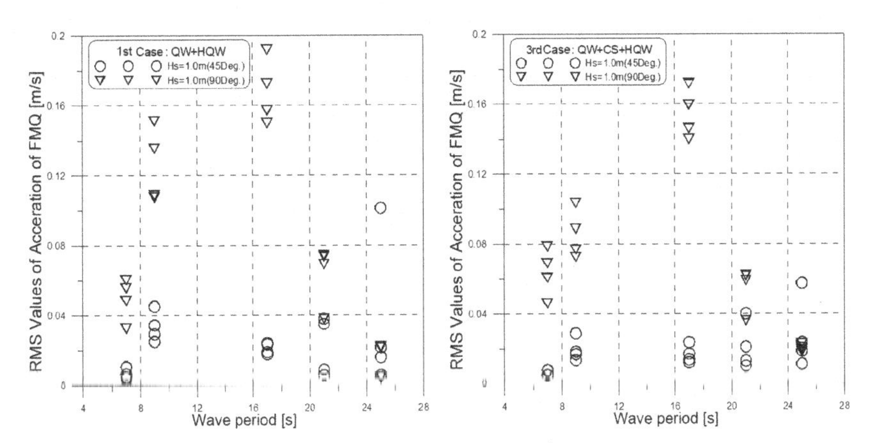

Fig. 4. Motion responses of FMQ and CS (1st and 3rd cases, wave dir. 45 degree).

The maximum surge and sway motions of FMQ in irregular waves are shown in Fig. 5, which are from numerical and hydraulic model tests. Thick line means allowable motion criteria for a large ship suggested by PIANC. The motion responses of FMQ tend to become larger as wave period gets longer and are below the allowable motion criteria at wave period of 7 seconds.

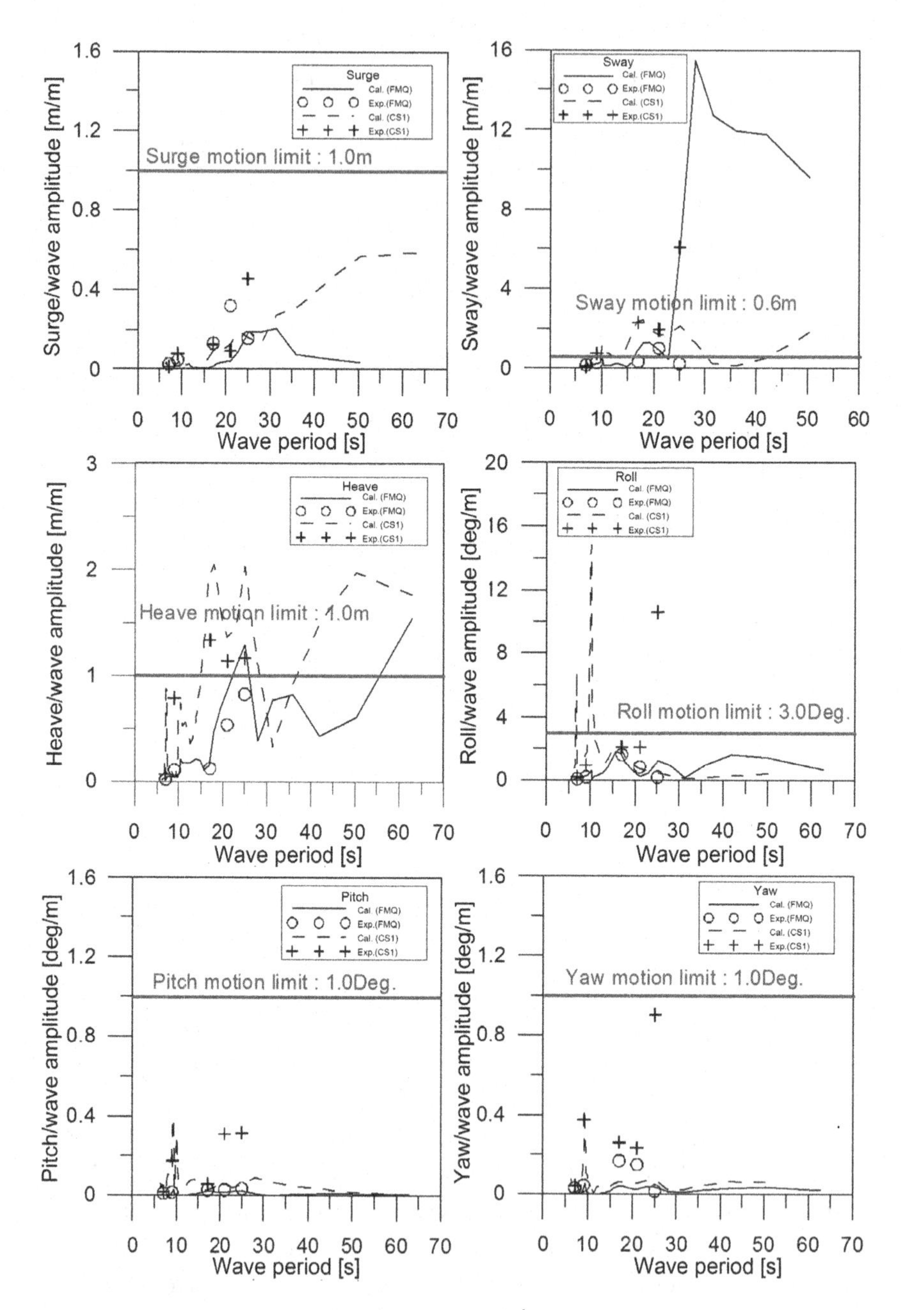

Fig. 5. Motion responses of FMQ and CS (3rd case, wave dir. 90 degree).

When the wave attack angle is 90 degrees the acceleration response is largest at 17 seconds and when it is 45 degrees the response is relatively large (Fig. 6). When FMQ is berthing and leaving a land based quay the horizontal and vertical accelerations of the crane are less than 0.2 m/s^2. Thus the crane on FMQ is considered to be stable.

Accelerations of container crane response to the irregular waves in the model test are shown in Fig. 7. They appear to be large when the waves are high and have long period. The accelerations are less than 0.13 m/s^2 and thus satisfy the Mega-float's criteria.

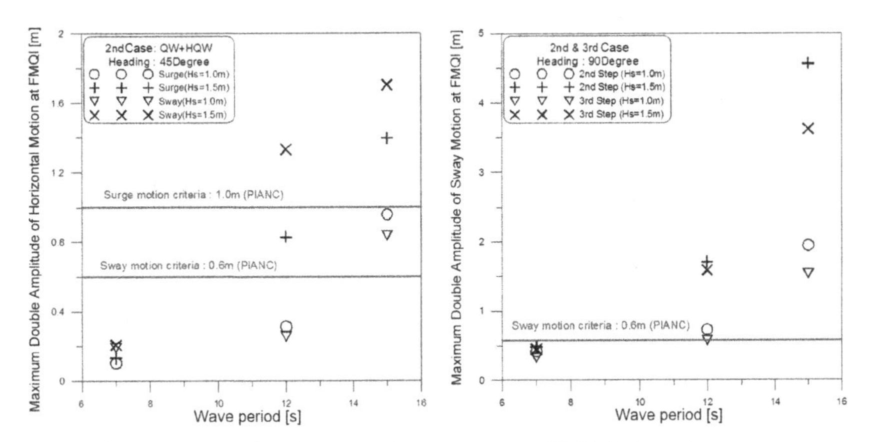

Fig. 6. Horizontal and sway motion responses of FMQ in irregular waves.

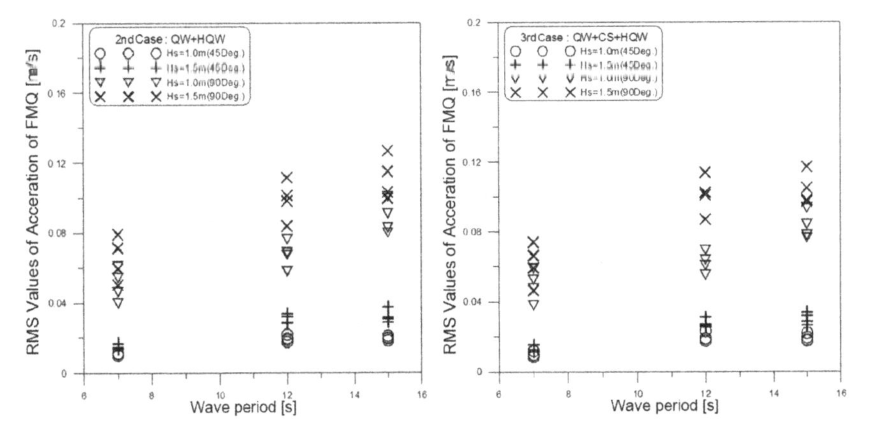

Fig. 7. Acceleration responses of CC at a container ship in irregular waves.

Crane Productivity of FMQ

In order to evaluate crane productivity of FMQ a series of test has been used using a crane simulator. 22 row-quay crane (CC) installed at Busan New Port and 6,300 TEU class container ship are chosen for the test. It is assumed that four CCs are used for unloading and another four CCs are for loading, and hatch cover is to be moved two times.

Five CC operators working at Busan Port were temporally employed for loading and unloading simulation tests. As FMQ is a movable type floating structure, winds and waves might cause some effects on its movement and further crane operation at FMQ. With input data of the FMQ motion for the crane simulator workability and productivity of the crane were evaluated. Those data were given from the hydraulic model tests after coordinate transformation process.

Table 3 shows simulation scenarios for the operation conditions of the cranes on land based fixed quay (QW) and FMQ with respect to wind and waves. This will be used for the evaluation of crane workability and productivity due to the motion of quays when they are in normal and operation limit conditions.

Table 3. Simulation scenarios for productivity tests of the cranes on QW and FMQ in normal and operation limit conditions with respect to wind and waves.

Scenario		Fixed	Normal operation condition	Operation limit condition
QW	CS (H;T)	0 ; 0	0.2 ; 7	0.7 ; 12
	Quay(H;T)	0 ; 0	0 ; 0	0 ; 0
	Wind	0 m/s	3.7 m/s	16 m/s
FMQ	CS(H;T)	0 ; 0	0.2 ; 7	0.7 ; 12
	Quay(H;T)	0 ; 0	0.2 ; 7	0.7 ; 12
	Wind	0 m/s	3.7 m/s	16 m/s

H : wave height (m), T : wave period (s).

Simulation results of crane productivity by five professional crane operators for six scenarios are shown in Table 4. The operators are B, C, D, E, F whose skills are different from excellent to fair. When the quay is fixed the productivity of crane on FMQ is 93.2% in efficiency of that of CC on QW. On the normal operation condition it is 98.9% and on the operation limit condition 87.5% in efficiency. In general the efficiency in terms of crane productivity on FMQ is appeared to be over 93% of land based quay crane.

Table 4. Productivity through crane simulation tests by CC operators.

(unit : VAN)

Scenario	Fixed					Normal operation condition					Operation limit condition				
Operator	B	C	D	E	F	B	C	D	E	F	B	C	D	E	F
QW	32.7	36.5	38.7	38.2	35.6	36.5	39.4	39.0	37.1	34.4	32.4	35.9	32.7	29.5	29.2
Mean productivity	36.3					37.3					31.9				
FMQ	31.9	38.4	32.7	33.8	32.6	39.8	40.6	32.8	35.5	36.0	34.2	22.3	27.5	27.1	28.6
Mean productivity	33.9					36.9					27.9				
FMQ efficiency	93.3%					98.9%					87.5%				
Avg. efficiency	93.2%														

2.3. *Concluding Remarks*

It is shown that FMQ sufficiently satisfy the stability and workability requirements for crane operations through numerical and hydraulic model experiments. Productivity of the crane on FMQ is over 93% in comparison with that of land based fixed quay when they are in fixed, normal and operation limit conditions.

3. Application of FMQ for the West Terminal of Busan New Port

3.1. *West Terminal adopting FMQ*

To examine the applicability and feasibility of the FMQ, the system was applied to the west terminal of Busan New Port (refer to Fig. 8) of Korea. As shown in Fig. 1, the both side loading/unloading system was adopted by using the FMQ with the dolphin mooring system. Five gantry cranes are located at the landside quay with an outreach of 63 m, a back reach of 20 m and a lift height of 58 m, which can handle 22 TEU wide container ships.

Table 5. Dimensions of FMQ and container ships.

Type	Length (m)	Width (m)	Depth (m)	Draft (m)	K_G (m)
FMQ	480.0	160.0	8.0	6.0	6.14
VLCS (15,000 TEU)	400.0	57.5	16.0	14.0	23.2
CS (4,500 TEU)	250.0	36.3	15.0	13.6	13.6

Four same cranes are on the FMQ for both side loading/unloading. Four cranes are arranged on the outside of the FMQ, which can serve 13 TEU wide feeder ships. For transferring and storing/handling containers on container yard (CY) and FMQ, yard tractors (YT's) and rail mounted gantry cranes (RMGC's) and are used. Table 5 shows the dimensions of the FMQ and VLCS of 15,000 and CS of 4,500 TEU considered, and the water depth in front of the fixed quay is 16m. For a VLCS of 15,000 TEU FMQ is to be used for both sides loading and unloading in combination with land quay. However for CS of 4,500 TEU FMQ is to be apart 300 m from land quay and used for two ships moored at FMQ and land quay, respectively (Fig. 9c and 9d).

Fig. 8. FMQ development site in Busan New Port.

3.2. *Feasibility Analysis*

Stability in Operational and Extreme Conditions

In order to analyze safety, stability and workability of FMQ for Busan New Port a series of hydraulic model test was made in a scale of 1/100. Models of FMQ and VLCS of 15,000 TEU and CS of 4,500 TEU were made in which their dimensions are shown in Table 5. Measuring instruments are same as those used in Section 2.2.

The model test has been carried out for 8 cases as shown in Fig. 9, which are operational limit conditions of $H = 0.7$ m, 1.0 m with $T = 5$ s ~ 30 s and survival conditions of $H_s = 2.8$ m wave height with $T_p = 15.5$ s (refer Table 6).

Table 6. Wave conditions for model test.

Wave	Regular		Irregular
Wave Height (m)	0.7	1.0	2.8
Wave Period (s)	5, 7, 9, 11, 13, 15	20, 25, 30	15.5

Dynamic response signals of the VLCS and CS moored at FMQ are shown in Fig. 10. Operational Limit of container ships is drawn in a dotted line and FMQ fender force limit in a solid line. When T is smaller than 15s both FMQ and VLCS's responses for six degree of freedom are in a similar pattern, but as wave period increases the response of VLCS gets relatively larger than that of FMQ. Within the range of $T < 15$ s surge and sway motions are below the allowable limit, and sway motion is larger than surge. This means that waves propagate in 45 degrees and exert dynamic forces to a longer sectional area of FMQ. Heave motions are below the allowable limit except for $T = 30$ s and yaw motions are below the limit for $T = 5$ s ~ 30 s.

Maximum forces on dolphin fenders of FMQ are shown in Fig. 11. Allowable maximum fender forces are resulted in 152 tons when the distortion rate is 15% for the fender of height 1.6 m. Fender forces are also increased with similar pattern to FMQ responses depending on wave periods, but they are within the limit when T is smaller than 15 s.

(a) Case 1 : QW

(b) Case 2 : QW+FMQ

(c) Case 3 : QW+HQW+VLCS

(d) Case 4 : QW+CS+FMQ

(e) Case 5 : QW+CS+FMQ

(f) Case 6 : QW+FMQ

(g) Case 7 : QW+FMQ

(h) Case 8 : QW+CS

Fig. 9. Configuration of 8 cases for hydraulic model tests for the FMQ system.

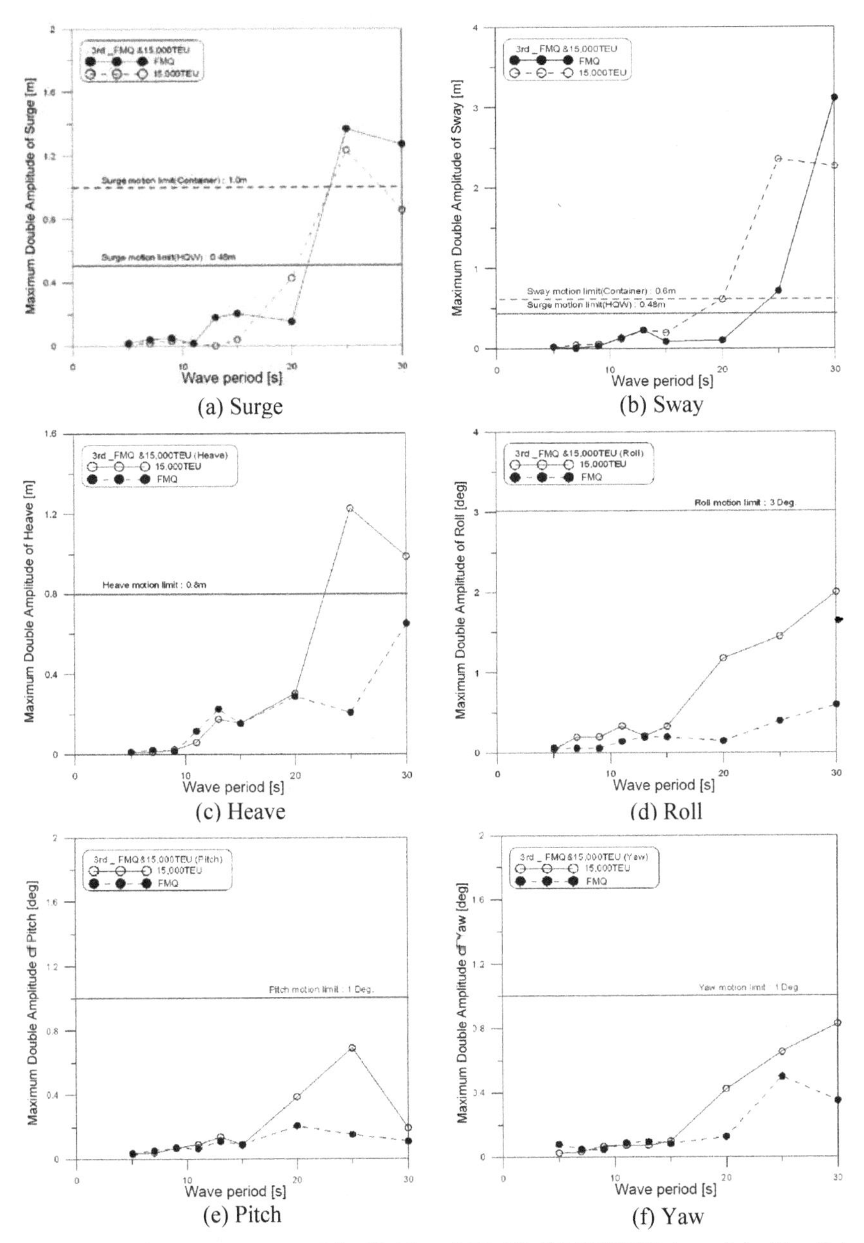

(a) Surge (b) Sway

(c) Heave (d) Roll

(e) Pitch (f) Yaw

Fig. 10. Maximum responses of the FMQ and VLCS (15,000TEU) (case 3 in Fig. 5) in regular wave conditions.

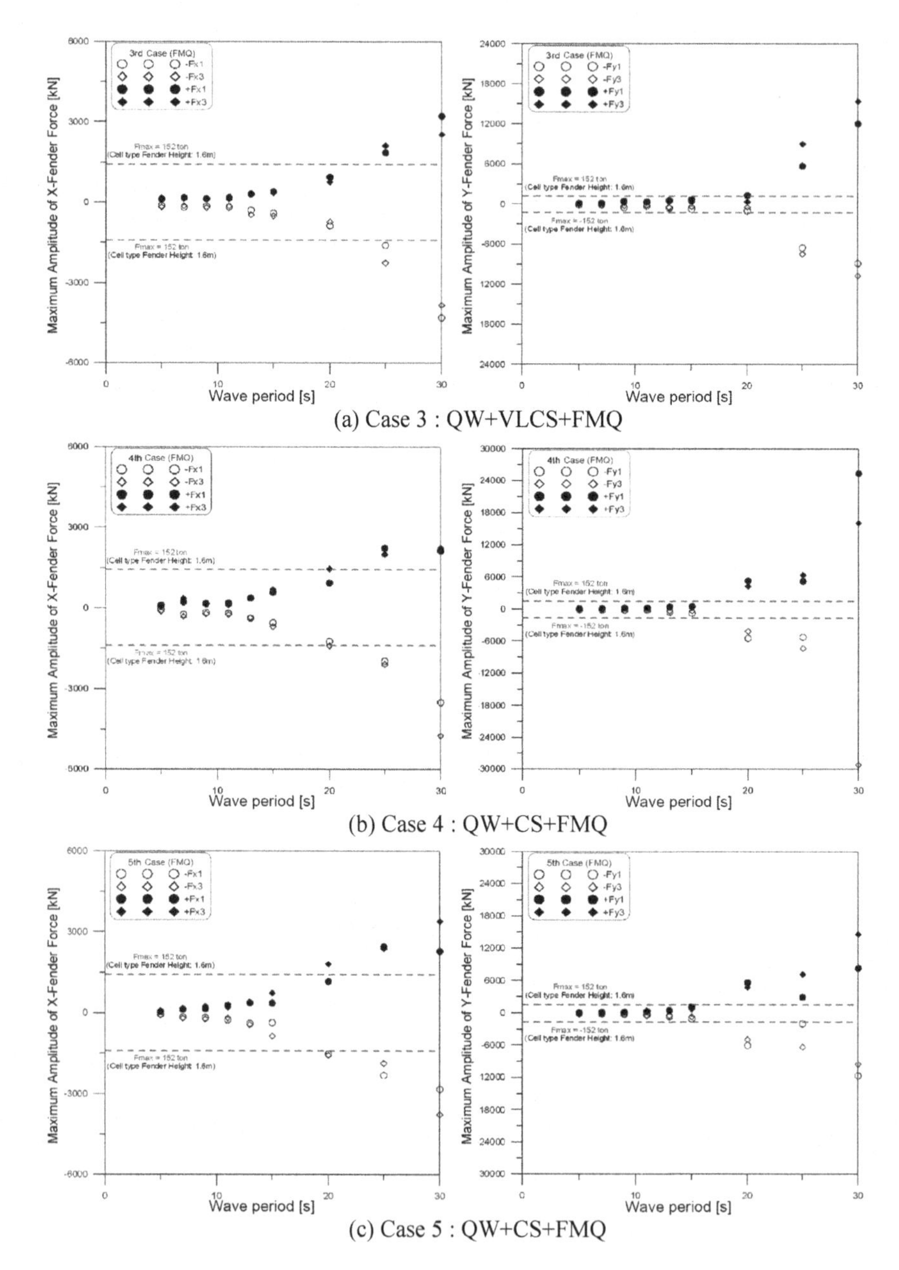

(a) Case 3 : QW+VLCS+FMQ

(b) Case 4 : QW+CS+FMQ

(c) Case 5 : QW+CS+FMQ

Fig. 11. Maximum forces on dolphin fenders of the FMQ (case 3, 4, 5 in Fig. 5) in regular wave conditions.

Maximum acceleration of the container crane (CC) on FMQ are measured and shown in Fig. 12. In general the acceleration is an important factor on cargo weight, equipment, and passenger's seasick. They are within the allowable limits of container loading/unloading condition

However there is no guideline for the limits on the acceleration of loading/unloading CC on FMQ, even on a marine structure moored in a harbor, and also for the acceleration limit of CC operator's condition. In this study a perceptional limit of acceleration is used for the analysis, which is same as that suggested from the Mega float study in Japan. They are 0.16 m/s^2 (RMS) for motion period 5 s and 0.32 m/s^2 (RMS) for 10 s. It appears that both horizontal and vertical accelerations of CC on FMQ are less than 0.1 m/s^2 in cases of with or without container ships

In Fig. 13, exceedance probability of fender and mooring line forces is shown when design wind blows from sea to land. Fender forces are within the range of maximum forces except for Fd1 case. Mooring line force is less than 0.1% of its maximum. From this results the fender and rope forces are in part little bit larger than initial linear increasing limit, but the fender force in breaking condition reaches 250 ton which is 1.68 times the maximum. Thus it might be mentioned that FMQ moored at land berth may be safe even in the survival condition.

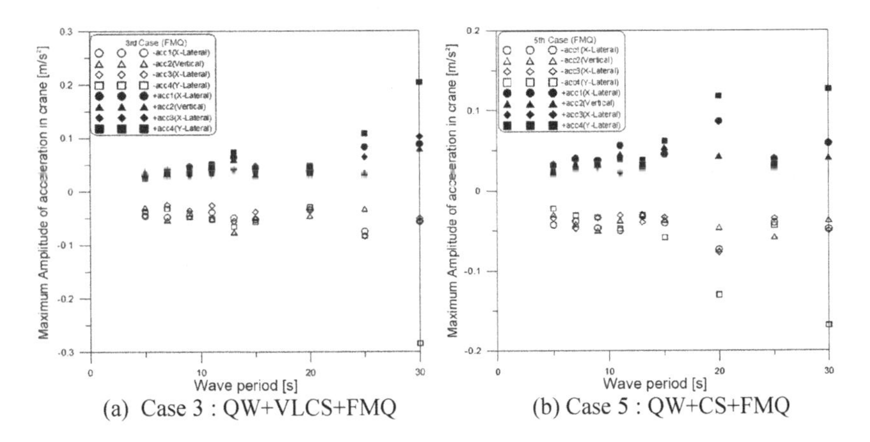

(a) Case 3 : QW+VLCS+FMQ (b) Case 5 : QW+CS+FMQ

Fig. 12. Maximum acceleration of the crane on the FMQ (case 3, 5 in Fig. 5) in regular wave conditions.

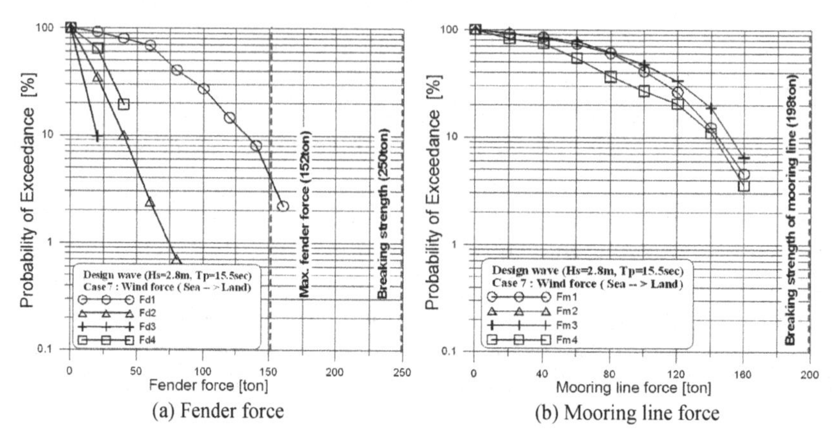

(a) Fender force (b) Mooring line force

Fig. 13. Exceedance probability of fender and mooring line forces of the case 7 under the design wave condition.

Economic Feasibility

The FMQ can increase the loading/unloading capacity of a container ship (CS) by adding the extra number of quay cranes (CC) on the water side. And thus the berth productivity can certainly be increased. The FMQ for a 15,000 TEU class CS can handle 746,967 TEU/year using four quay cranes for the large ship and two smaller quay cranes for 2,000 TEU feeder vessels. The productivity of the cranes is based on those of the Busan Port. When it is used with a land based conventional berth (QW) as the proposed west terminal of the port, the berth productivity will be 1,413,632 TEU that is 212% increase comparing with the present QW productivity of 666,665TEU with five quay cranes. It is assumed that the containers are transported from the quay to container storage yard (CY) by using YT and stored/handled by RMGC. The productivity of a mobile FMQ for a 4,000 TEU class is to be 1,813,634 TEU (151% increase) when it is used with existing land based three QWs. This means that the FMQ provides extra productivity and transshipment handling capacity when using with QWs.

Berth productivity is also analyzed in terms of ship residence time as shown in Table 7. In case of 8,000 TEU with FMQ ship productivity of fixed (land) quay is 120 TEU/hr with 6 CC and 64 TEU/hr with four CCs, and operational time of loading and unloading will be reduced by 30%

comparing with the case of w/o FMQ. As the ship residence time of 15,000 TEU container ship at the quay can be reduced by 27% and thus the ship also can be served within 24 hours.

The investment costs of construction and equipment, and the operational costs of maintenance, labor and energy etc. were estimated for the FMQ and QWs. The benefits are also estimated. The B/C ratio of the FMQ is calculated as 1.13 for the west terminal which has one berth for FMQ operated at the end of the terminal and four other berths (Table 8). Even though the productivity greatly depends on logistics and terminal operation / FMQ workability etc., this means that the FMQ is feasible economically.

Table 7. Productivity in terms of the ship residence time.

CS Size (TEU)	Productivity(ea/hour)		Operation time (hour)		Reduced time	
	QW	FMQ	w/o FMQ (A)	w/ FMQ (B)	Hours (A-B)	% [(A-B)/A *100]
3,000 ~ 4,000	66 (3x22)	52.8 (3x22x0.8)	17.6	9.8	7.8	44%
4,000 ~ 5,000	92 (4x23)	73.6 (4x23x0.8)	12	6.6	5.4	45%
5,000 ~ 7,000	100 (5x20)	64 (4x20x0.8)	16.7	10.2	6.5	39%
8,000	120 (6x20)	64 (4x20x0.8)	16.7	12.2	4.5	30%
10,000	140 (7x20)	64 (4x20x0.8)	19	13	6	31%
15,000	180 (9x20)	64 (4x20x0.8)	22	16	6	27%

Table 8. Economic feasibility.

Total Cost (construction, equipment, labor, power)	1.83 billion USD
Total Benefit (anchorage, loading/unloading)	2.07 billion USD
NPV (Net Present Value)	0.24 billion USD
B/C Ratio (Benefit Cost Ratio)	1.13
IRR (Internal Rate of Return)	8.07%

3.3. *Concluding Remarks*

When FMQ is applied to Busan New Port it is appeared that FMQ is safe even in a design wave condition of 50 year-return period and also sufficiently workable in the operation limit condition. From the economic feasibility analysis B/C was calculated as 1.13. Therefore FMQ application to Busan New Port seems to be feasible both in technical and economic aspects.

4. Conclusions and Outlook

A new berth system with a floating mobile quay (FMQ) is suggested to greatly enhance loading and unloading capability of the existing land based berth in a hub port. The FMQ is basically consisted of a very large pontoon type floating structure with azimuth thrusters controlled by the dynamic positioning system, mooring systems, and access bridges to the land side berth.

Applying the adjustable type FMQ to the west terminal of Busan New Port as a hub port in the North-East Asia, it is shown that the berth system with a floating quay is feasible, technically as well as economically even for a VLCS in 15,000 TEU class.

For practical use of the mobile type FMQ in a conventional hub port, further detail analysis of the system has recently been carried out[11]. The analysis includes FMQ maneuvering and fast transferring techniques, mooring technique and high efficient container handling equipment on the floating quay and berth/terminal operating system. A FMQ test model (50 m x 30 m x 5 m RC consists of 4 modules) was fabricated in a dock yard and the modules are installed / connected in situ.

A field model test and maintenance monitoring of the floating quay system have been made[12]. Based on the measured motion and mooring line data of the model FMQ, the stability/safety of the FMQ was checked, and also a quay crane simulator was operated in order to estimate it's productivity. From the financial analysis of the terminal having three conventional berths with the FMQ for serving both super and conventional container ships, the terminal with FMQ appears to be more beneficial than that without the FMQ.

Acknowledgments

This research was fully supported by the Korea Ministry of Land, Transport and Maritime Affairs through Project entitled "Research & Development of New Harbor Technology (Hybrid Quay Wall) for a Very Large Container Ship (VLCS)".

References

1. Yap, W.Y. and J.S.L. Lam, 80 million-twenty-foot-equivalent-unit container port ? Sustainability issues in port and coastal development. *Ocean & Coastal management,* 71(2013).
2. B. A. Pielage and J. C. Rijsenbrij, *World Port Development,* Oct. (2005).
3. C. M. Wang, Wu, T. Y., Choo, Y. S., Ang. K. K., Toh, A. C., Mao, W. Y. and A. M. Hee, *Marine Structures*, 19 (2006).
4. Ohakawa, Yutaka, Proceeding of International workshop on Very Large Floating Structures, VLFS' 96 (1996).
5. E. Isobe, Proceeding of International workshop on Very Large Floating Structures, VLFS' 99 (1999).
6. H. Suzuki, *Marine Structures,* 18(2) (2005).
7. Keith R. McAllister, Proceeding of International workshop on Very Large Floating Structures, VLFS' 96 (1996).
8. G. Remmers, Proceeding of International workshop on Very Large Floating Structures, VLFS' 99 (1999).
9. P. Palo, *Marine Structures,* 18(2) (2005).
10. J-W. Chae, Park, W-S., Yi, J-H. and Jeong, G-I., *Proc. of ICCE2008 Hamburg.* 3899-3911 (2008).
11. Ministry of Land Transport and Maritime, final report of research and development of new harbor technology (Hybrid Quay Wall) for a very large container ship. (2010) (in Korean).
12. Li, J. K., J. H. Kim, W. M. Jeong, and J. W. Chae, Vision-based dynamic motion measurement of a floating structure using multiple targets under wave loadings. *J. Korean Society of Civil Engineers,* 32(1A) (2012) (in Korean).

CHAPTER 7

SURROGATE MODELING FOR HURRICANE WAVE AND INUNDATION PREDICTION

Jane McKee Smith

US Army Engineer Research and Development Center, Coastal and Hydraulics Laboratory, 3909 Halls Ferry Road, Vicksburg, Mississippi, USA, 39180
E-mail: jane.m.smith@usace.army.mil

Alexandros A. Taflanidis and Andrew B. Kennedy

Department of Civil and Environmental Engineering and Earth Sciences, University of Notre Dame, Notre Dame, Indiana, USA, 46556
E-mail: a.taflanidis@nd.edu, andrew.kennedy@nd.edu

Real-time forecasting of hurricane waves and inundation has required a difficult choice between speed and accuracy. High-fidelity, high-resolution numerical simulations provide accurate deterministic simulations. These simulations are computationally expensive, so only a limited number of runs can be made for real-time forecasts, and it is difficult to account for error in forecast parameters or model simulations. Simulations with lower resolution and fidelity models can be made more quickly, facilitating probabilistic simulations (hundreds of Monte Carlo simulations that sample the probability space representing the forecast error). However, low-fidelity, low-resolution simulations can result in large errors due to neglected processes or under-resolved environment (either over or under prediction of waves, surge, and inundation). An alternate approach is to create a forecast system that is fast and robust by applying a surrogate modeling method. Waves and inundation are precomputed based on the high-fidelity, high-resolution models. A surrogate model is then developed based on the database of high-fidelity simulations, offering satisfactory accuracy and enhanced computational efficiency (evaluation of responses within a few seconds). The efficiency of the surrogate model allows both deterministic and probabilistic simulations in seconds to minutes on a personal computer.

1. Introduction

Hurricanes, typhoons, and tropical cyclones have resulted in high death tolls and devastating damage in many parts of the world. Prior to hurricane forecasting in the US, landfalling storms resulted in large death tolls, e.g., the 1900 Galveston Hurricane and the 1928 Lake Okeechobee Hurricane resulted in 8000 and 2500 deaths[1], respectively. In 1970, Tropical Cyclone Bhola in the Ganges Delta caused an estimated 500,000 deaths, and other western Pacific and Indian Ocean storms have resulted in death tolls greater than 100,000. In the US, even recent storms have resulted in large damages and death tolls, e.g., Hurricane Katrina in 2005 caused an estimated $108 billion in damages and 1200 deaths[1] and Sandy in 2012 caused greater than $50 billion in damages and 72 deaths[2]. Hurricane Sandy was downgraded to a post-tropical cyclone before landfall, but the impacts were large because the storm influenced a long section of the US coast with a high population density. Damages and deaths relate not only to the strength and size of the storm, but also the physical parameters of the coastal region, population density, and infrastructure vulnerability.

The destructive Atlantic hurricane seasons of 2004, 2005 and 2008 increased awareness of hurricane risk in the US. Following Hurricane Katrina in 2005, a significant effort was put forth to validate high-fidelity, high-resolution, coupled surge and wave models to quantify and improve model performance[3,4,5,6,7]. Skill indices (normalized root-mean-square error) for the surge modeling range from 0.1 to 0.3 and for wave modeling range from 0.2 to 0.4. The larger errors tend to be in shallow, inshore areas that are most sensitive to local water depth and roughness. Model accuracy and output resolution make these powerful tools for assessing flood risk. The downside is that computational cost can exceed thousands of CPU hours for each storm. Thus, even though the models are optimized for parallel computing on thousands of processors, the run times are too long to execute hundreds or thousands of runs in real time for probabilistic forecasts. Quality control of the large model output is also difficult in real time.

This realization has incentivized researchers to investigate interpolation and surrogate modeling methodologies[8,9] that use information from existing databases of high-fidelity simulations, to efficiently approximate hurricane wave and inundation responses. This use of surrogate models has been further motivated by the fact that various such databases are constantly created and updated for regional flooding and coastal hazard studies[10]. The underlying assumption is that hurricane scenarios may be represented by a simplified parametric description with a small number of model parameters. These parameters typically correspond to the characteristics close to landfall - an idea motivated ultimately by the Joint Probability Method (JPM)[11] which is utilized widely for hurricane risk assessment.

This chapter discusses an efficient framework for establishing surrogate models (also referenced as metamodels) to predict hurricane wind, surge, and wave responses. It discusses details of the high-fidelity modeling for establishing the initial database, but primarily focuses on the development of a kriging surrogate model as well as its implementation for highly efficient real-time risk assessment. Illustrative examples are provided for Hawaii and for New Orleans, Louisiana. Section 2 of this chapter describes the high-fidelity, high-resolution surge and wave models, and Section 3 describes surrogate modeling. Section 4 provides examples of surrogate models developed for Hawaii and for New Orleans, focusing on implementation for real-time risk assessment. Section 5 provides a summary and discusses future work.

2. Numerical Modeling of Hurricane Waves and Surge

Surrogate models are established by exploiting information in existing databases of numerical simulations. Ultimately, they allow the adoption of high-fidelity models that can facilitate highly accurate predictions of hurricane responses, with the only consideration being the upfront cost for establishing these databases. This section provides brief details on the numerical models used in the context of the case studies discussed later. The ADvanced CIRCulation model (ADCIRC) was utilized for inundation predictions and the Simulating WAves Nearshore (SWAN) model, running in parallel with ADCIRC, was used for wave predictions.

The ADCIRC model solves the forced, depth-integrated shallow water equations over unstructured triangular elements to give estimates of surge elevations and velocities over the course of a storm[5,12]. Typically, high-resolution grids feature element sizes of several km in the deep ocean, scaling down to 20-30 m in highly resolved nearshore regions. It is highly-parallelized to run on thousands of computational cores, so surge for these very high resolutions may be computed relatively efficiently on high-performance computing systems.

The SWAN model solves the spectrally-averaged equations of wave motion[13] over the same high-resolution unstructured triangular grids as ADCIRC, ensuring that interpolation errors between the models are minimized. Simulations have two-way interaction between the two models, with ADCIRC-computed currents and water levels affecting SWAN wave propagation, while SWAN-derived gradients in radiation stresses force ADCIRC currents and water levels in shallow regions. Both models are forced by identical wind fields and use the same atmospheric drag laws, ensuring inter-model consistency.

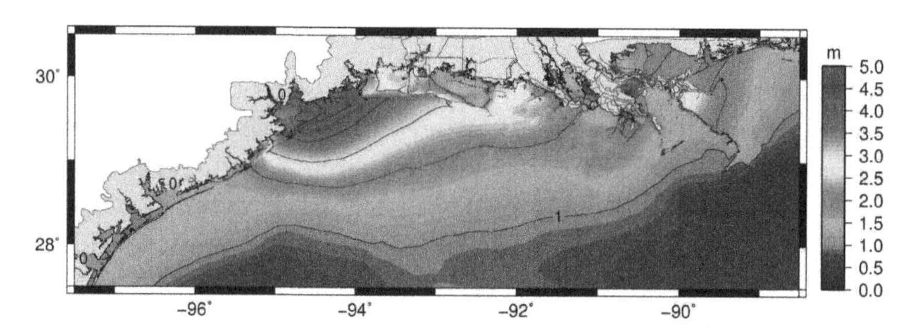

Fig. 1. Maximum SWAN+ADCIRC surge elevation (NAVD88) computed during Hurricane Ike.

These models have been widely used in recent years to hindcast large storms using the best available wind fields[7], and for prognostic purposes using synthetic data to examine potential storm scenarios[6]. Figure 1 shows computed maximum water levels during Hurricane Ike in the Gulf of Mexico, and Fig. 2 shows the computed maximum wave height during the same storm. Figure 3 gives a comparison between computed and measured High Water Marks (HWM)[7]. This and many other storm comparisons have shown good accuracy for both waves and surge[6,12]. Inter-model and inter-grid comparisons detailed in Kerr *et al.*[12,14] suggest that there is a continuous improvement in simulation accuracy as resolution increases in nearshore regions and on flooded land. Thus, the high resolutions used here for prognostic studies give good confidence that simulations may be relied upon to give good accuracy in evaluating the consequences of storms yet unformed, and will provide a useful tool both for long-term planning studies and for shorter-term decision-making during an approaching storm.

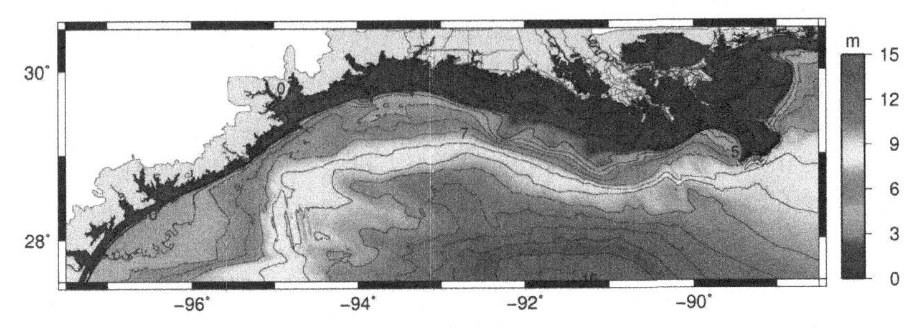

Fig. 2. Maximum SWAN+ADCIRC significant wave height computed during Hurricane Ike.

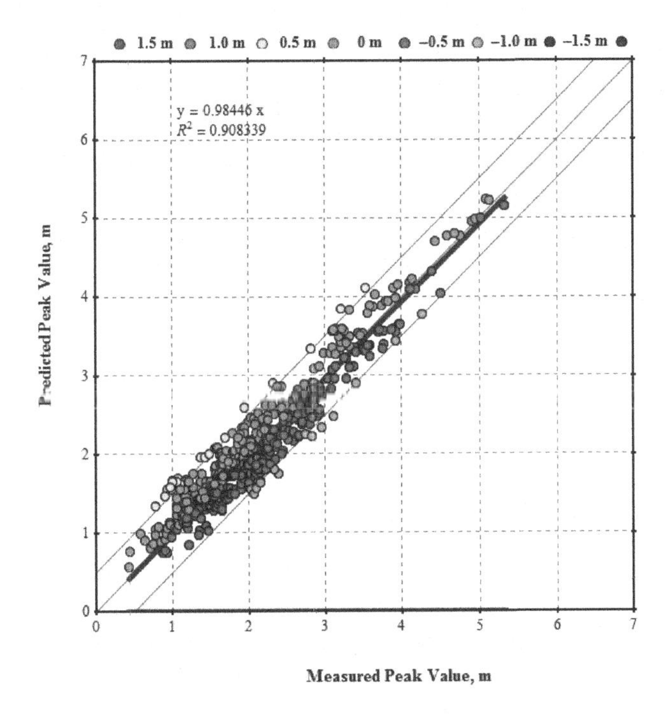

Fig. 3. Comparison between measured high water marks (HWM) during Hurricane Ike and computed values using SWAN+ADCIRC.

When implemented for surrogate modeling purposes, these high-fidelity models are run for hundreds of representative storms to build up a database of wave heights and surge elevations arising from given storm conditions. These representative storms should cover the entire range of hurricane scenarios for which the surrogate models are expected to be required to provide predictions, to avoid the requirement of extrapolation.

3. Surrogate Modeling

Surrogate models can be developed by utilizing the information in existing databases, with the latter developed either explicitly for these purposes (i.e., support surrogate modeling for real-time risk assessment) or under other motivation (regional flood studies, for example). The specific surrogate model promoted here is kriging. This preference is because (i) it is highly efficient and robust as it relies completely on matrix manipulations without requiring any matrix inversions at the implementation stage (the latter could be a problem with moving least squares response surface approximations that share the rest of the advantages of kriging and has been also used for hurricane response predictions), (ii) it can provide simultaneously predictions for different type of responses (inundation and wave predictions) over an extensive coastal region, and (iii) it offers a generalized approach that is easily extendable to different regions (meaning different databases).

3.1. *Hurricane modeling*

Similar to JPM, the surrogate modeling implementation requires each hurricane event be approximated by a small number of model parameters, corresponding to its characteristics at landfall, such as (i) a reference landfall location x_o, (ii) the angle of approach θ, (iii) the central pressure c_p, (iv) the forward speed v_f, (v) the radius of maximum winds R_m, (vi) the Holland number B, or (vii) the tide characteristics e_t. The exact selection depends on the applications considered, but ultimately defines the n_x dimensional vector of input model parameters $\mathbf{x}$. In the case studies discussed in Sec. 4, the model parameter vector is taken as $\mathbf{x} = [x_o\ \theta\ c_p\ v_f\ R_m]$. The variability of the hurricane prior to landfall is of course important, and within this framework it is approximately addressed by appropriate selection of hurricane track history prior to landfall when creating the initial database, so that important anticipated variations, based on historical data, are efficiently described. Figure 4 shows the tracks considered for Hawaii in the study by Kennedy *et al.*[10].

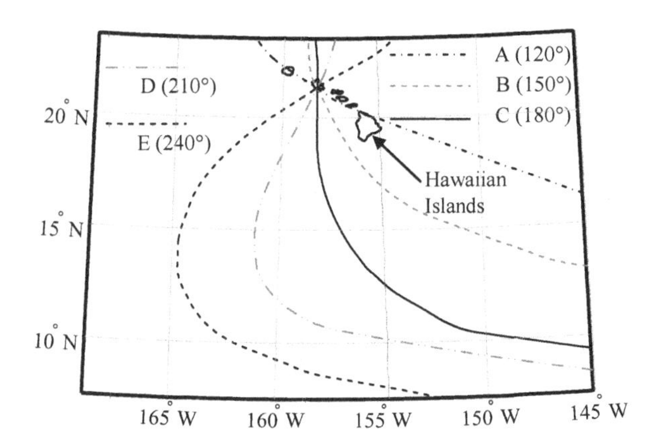

Fig. 4. Basic storm tracks (A-E) considered in the study[10]. In parenthesis the angle of final approach θ (clockwise from North) is indicated.

Let, then, $\mathbf{y}(\mathbf{x})$ denote the output vector with dimension n_y for a given input $\mathbf{x}$ (a specific hurricane scenario). The kth component of this vector is denoted by $y_k(\mathbf{x})$ and pertains to a specific response variable (such as storm inundation or significant wave height) for a specific coastal location and a specific time instance. The augmentation of these responses for all different locations in the region of interest and for all different time instances during the hurricane history ultimately provides the n_y-dimensional vector $\mathbf{y}(\mathbf{x})$. Note that the dimension of n_y can easily exceed 10^6, depending on the type of application. Ultimately we have an available database of n high-fidelity simulations, with evaluations of the response vector $\{\mathbf{y}^h; h=1,\ldots,n\}$ for different hurricane scenarios $\{\mathbf{x}^h; h=1,\ldots, n\}$. We will denote by $\mathbf{X}$ and $\mathbf{Y}$ the corresponding input and output matrices respectively with $\mathbf{X}=[\mathbf{x}^1 \ldots \mathbf{x}^n]$ and $\mathbf{Y}=[y(\mathbf{x}^1) \ldots y(\mathbf{x}^n)]$. These high-fidelity simulation outputs (database) are referenced also as the *training set* or *support points*.

Before developing the surrogate model, a correction is required for the inundation for any inland locations where we are interested in providing predictions. The challenge with these locations is that they do not *always* get inundated (in other words dry locations might remain dry for some or even most storms). Though some scenarios in the database may include full information for the storm surge (when the location does get inundated), some provide *only* the information that the location remained dry. Thus, the initial database needs to be corrected so that each scenario in our database can provide full, exact or approximate, information. This can be established through the approach proposed in Taflanidis *et al.*[15] The storm elevation is described with respect to the mean sea level as a reference point. When a location remains dry, the incomplete information is resolved by selecting (approximating) as its storm elevation the one corresponding to the nearest location (nearest node in our high-fidelity numerical model) that was inundated. Once the database is adjusted for the scenarios in which each location remained dry, the updated response database is used for building the surrogate model. Finally, comparison of the predicted surge elevation to the ground elevation of the location provides the answer as to whether the location was inundated or not, whereas the inundation depth is calculated by subtracting these two quantities. Thus, this approach allows us to gather simultaneous information about both the inundation (binary answer; yes or no) as well as the inundation depth, although it does involve the aforementioned approximation for enhancing the database with complete information for the surge elevation for all hurricane scenarios.

3.2. *Principal component analysis*

For applications over large coastal regions, a very large number of prediction points are required to compute flood risk. The efficiency of the surrogate modelling approach can be then enhanced by coupling it with a dimensional reduction technique. This technique exploits the strong potential correlation between responses at different times or locations in the same coastal region (which are the main attributes contributing to the high dimension of the output vector) to extract a smaller number of latent outputs to represent the initial high-dimensional output. Note that this is not a necessary step, but it can provide significant savings for real-time applications with respect to both computational time as well as, and perhaps more importantly, with respect to the memory requirements[16], ultimately allowing the developed models to be implemented with limited computational resources.

Principal component analysis (PCA) is advocated here for this purpose. PCA starts by converting each of the output components into zero mean and unit variance under the statistics of the observation set through the linear transformation:

$$\underline{y}_k = \frac{y_k - \mu_k^y}{\sigma_k^y}, \text{ with } \mu_k^y = \frac{1}{n}\sum_{h=1}^{n} y_k^h, \sigma_k^y = \sqrt{\frac{1}{n}\sum_{h=1}^{n}\left(y_k^h - \mu_k^y\right)^2} \quad (1)$$

The corresponding (normalized) vector for the output is denoted by $\underline{\mathbf{y}}$ and the output matrix for the database by $\underline{\mathbf{Y}}$. The idea of PCA is to project the normalized $\underline{\mathbf{Y}}$ into a lower dimensional space by considering the eigenvalue problem for the associated covariance matrix $\underline{\mathbf{Y}}^T\underline{\mathbf{Y}}$ and retaining only the m_c largest eigenvalues. Then,

$$\underline{\mathbf{Y}}^T = \mathbf{P}\mathbf{Z}^T + \boldsymbol{\tau} \quad (2)$$

where $\mathbf{P}$ is the n_y x m_c dimension projection matrix containing the eigenvectors corresponding to the m_c largest eigenvalues, $\mathbf{Z}$ is the corresponding n x m_c output matrix for the principal components (latent outputs), and $\boldsymbol{\tau}$ is the error introduced by not considering all the eigenvalues[17]. The latent outputs, denoted by $\mathbf{z}_j$, j=1,…,m_c are the outputs with observations that correspond to the jth column of $\mathbf{Z}$. If λ_j is the jth largest eigenvalue, m_c can be selected so that the ratio

$$\sum_{j=1}^{m_c} \lambda_j \,/\, \sum_{j=1}^{n_y} \lambda_j \quad (3)$$

is greater than some chosen threshold r_0 (typically chosen as 99%). This then means that the selected latent outputs can account for at least r_0 % of the total variance of the data. It is then $m_c<\min(n, n_y)$ with m_c being usually a small fraction of $\min(n, n_y)$. For $n<<n_y$, obviously, $m_c<<n_y$, leading to a significant reduction of the dimension of the output. The relationship between the initial output vector and the vector of the latent outputs $\mathbf{z}=[z_1 \ldots z_{m_c}]$ is finally

$$\underline{\mathbf{y}} = \mathbf{P}\mathbf{z} \tag{4}$$

Note that PCA is completely data-driven and relies on the correlation of the output within database $\underline{\mathbf{Y}}$. It utilizes no information for the temporal or spatial correlation between the outputs at the different times/collations within the established grid used in the numerical high-fidelity simulations. When PCA is implemented, then for the surrogate modeling matrices $\mathbf{Z}$ with dimension n x m_c (providing the latent output information within the database) and $\mathbf{P}$ with dimension n_y x m_c (needed to transform from latent space back to original space) need to be kept in memory instead of the $\underline{\mathbf{Y}}$ with dimension n x n_y (providing the initial output information within the database). For most applications (when n_y is large) this leads to significant computational savings[16].

3.3. *Kriging metamodel*

Kriging is a highly efficient surrogate model. It corresponds to the best linear unbiased predictor[18], facilitates an interpolation (which means that when the new input is the same as one of the initial database points, the prediction will match the database high-fidelity response), and it not only gives the prediction but also the associated uncertainty, namely the local variance of the prediction error, which can be directly utilized in risk assessment as will be demonstrated later. Here the development of the metamodel is presented with respect to the latent output matrix $\mathbf{Z}$ obtained from PCA, though the approach is identical if PCA is not implemented (simply replace $\mathbf{Z}$ by $\mathbf{Y}$).

The fundamental building blocks of kriging are the n_p dimensional basis vector, $\mathbf{f}(\mathbf{x})$, and the correlation function $R(\mathbf{x}^j,\mathbf{x}^k)$. Typical selection for these two are full quadratic basis and generalized exponential correlation, respectively, leading to

$$\begin{aligned} &\mathbf{f}(\mathbf{x}) = [1 \;\; x_1 \;\; \cdots \;\; x_{n_x} \;\; x_1^2 \;\; x_1 x_2 \;\; \cdots \;\; x_{n_x}^2]; \quad n_p = (n_x+1)(n_x+2)/2 \\ &R(\mathbf{x}^j,\mathbf{x}^k) = \prod_{i=1}^{n_x} \exp[-\varphi_i \,|\, \mathbf{x}_i^j - \mathbf{x}_i^k \,|^{\eta}]; \quad \boldsymbol{\varphi} = [\varphi_1 \;\; \cdots \;\; \varphi_{n_x}] \end{aligned} \tag{5}$$

Note that the selection of correlation function influences the smoothness of the derived metamodel, and many other choices are available, i.e. cubic, spline, Matérn etc., among which Matérn gives greater flexibility as it has more parameters (or weights) to tune (optimize). For the set of n database points (defining the training set) with input matrix $\mathbf{X}$ and corresponding latent output matrix $\mathbf{Z}$, we define then the basis matrix $\mathbf{F}=[\mathbf{f}(\mathbf{x}^1) \ldots \mathbf{f}(\mathbf{x}^n)]^T$ and the correlation matrix $\mathbf{R}$ with the jk-element defined as $R(\mathbf{x}^j,\mathbf{x}^k), j,k=1, \ldots, n$. Also for every new input $\mathbf{x}$ we define the correlation vector $\mathbf{r}(\mathbf{x})=[R(\mathbf{x},\mathbf{x}^1) \ldots R(\mathbf{x},\mathbf{x}^n)]^T$ between the input and each of the elements of $\mathbf{X}$. The surrogate model approximation to $\mathbf{z}(\mathbf{x})$ is given by

$$\hat{\mathbf{z}}(\mathbf{x}) = \mathbf{f}(\mathbf{x})^T \boldsymbol{\alpha}^* + \mathbf{r}(\mathbf{x})^T \boldsymbol{\beta}^* \tag{6}$$

where matrices $\boldsymbol{\alpha}^*$ and $\boldsymbol{\beta}^*$ correspond to

$$\begin{aligned} \boldsymbol{\alpha}^* &= (\mathbf{F}^T\mathbf{R}^{-1}\mathbf{F})^{-1}\mathbf{F}^T\mathbf{R}^{-1}\mathbf{Z} \\ \boldsymbol{\beta}^* &= \mathbf{R}^{-1}(\mathbf{Z}-\mathbf{F}\boldsymbol{\alpha}^*) \end{aligned} \tag{7}$$

The prediction error for the jth output follows a Gaussian distribution with zero mean (unbiased predictions) and variance given by

$$\begin{aligned} &\hat{\phi}_j^2(\mathbf{x}) = \tilde{\sigma}_j^2[1+\mathbf{u}^T(\mathbf{F}^T\mathbf{R}^{-1}\mathbf{F})^{-1}\mathbf{u}-\mathbf{r}(\mathbf{x})^T\mathbf{R}^{-1}\mathbf{r}(\mathbf{x})], j=1,\cdots,m_c \\ &\text{where } \mathbf{u} = \mathbf{F}^T\mathbf{R}^{-1}\mathbf{r}(\mathbf{x})-\mathbf{f}(\mathbf{x}) \end{aligned} \tag{8}$$

and $\tilde{\sigma}_j^2$ corresponding to the diagonal elements of matrix

$$(\mathbf{Z}-\mathbf{F}\boldsymbol{\alpha}^*)^T\mathbf{R}^{-1}(\mathbf{Z}-\mathbf{F}\boldsymbol{\alpha}^*)/n \tag{9}$$

Through the proper tuning of the characteristics of the correlation function, $\boldsymbol{\varphi}$ and η, kriging can approximate very complex functions. The optimal selection of $[\boldsymbol{\varphi},\eta]$ is typically based on the Maximum Likelihood Estimation (MLE) principle, where the likelihood is defined as the probability of the n database points, and maximizing this likelihood with respect to $[\boldsymbol{\varphi},\eta]$ ultimately corresponds to the optimization problem

$$[\boldsymbol{\varphi},\eta]^* = \arg\min_{\boldsymbol{\varphi},\eta}\left[|\mathbf{R}|^{\frac{1}{n}}\sum_{j=1}^{n}\tilde{\sigma}_j^2\right] \tag{10}$$

where |.| stands for determinant of a matrix. This is an optimization problem that can be efficiently solved[19] even for problems with large dimensional n_x and large number of support points. Figure 5 presents an example for such a surrogate model implementation from the case study considered in Sec. 4; the maximum significant wave height in the vicinity of the Hawaiian Islands in response to a hurricane approximated in this case over a grid consisting of more than 10^4 points. The high-fidelity simulation for this application requires more than 2000 CPU hours whereas the kriging metamodel just 0.1 s. The accuracy of the predictions is evident from the comparison in this figure. This accuracy and remarkable computational efficiency of kriging, especially when combined with PCA for reduction of memory requirements for implementation in very large coastal regions[16], provides an ideal tool for fast storm inundation forecasting, whereas the approach is very robust because the high-fidelity model runs are quality controlled prior to the construction of the surrogate model. This tool can support real-time predictions for hurricane response and facilitate a probabilistic risk assessment by analyzing a large number of scenarios within a few minutes.

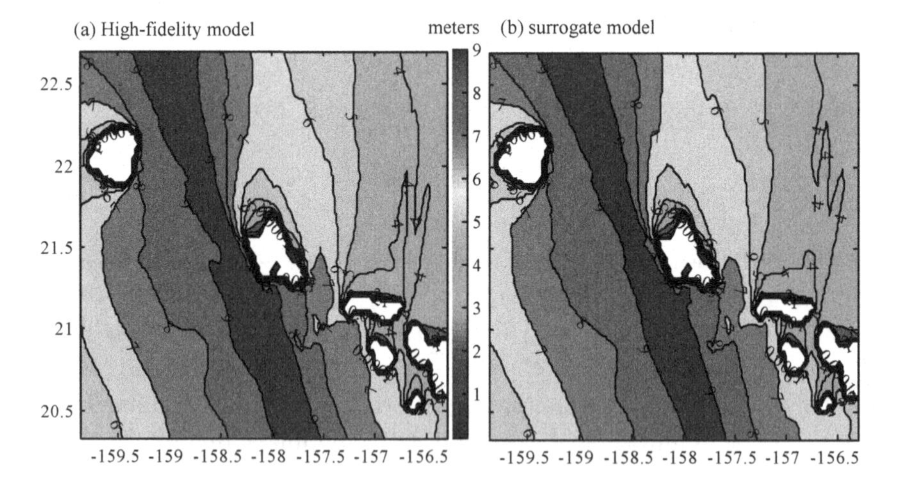

Fig. 5. Comparison between surrogate and high-fidelity model predictions for maximum significant wave height contours, for the case study discussed later in Sec. 4.

3.4. *Transformation to original space*

When the kriging surrogate modeling is coupled with PCA there is a need to transform the final predictions to the original space for the response. Since PCA is a linear projection, this is easy to establish: using (4), to transform the predictions from the latent space to the normalized response space, and then the inverse of (1) to transform this prediction back to the original space. The kriging prediction for $\mathbf{y}(\mathbf{x})$ is

$$\hat{\mathbf{y}}(\mathbf{x}) = \mathbf{\Sigma}^y \left(\mathbf{P}\hat{\mathbf{z}}(\mathbf{x})^T \right) + \boldsymbol{\mu}^y \tag{11}$$

where $\mathbf{\Sigma}^y$ is the diagonal matrix with elements σ_k^y, and $\boldsymbol{\mu}^y$ is the vector with elements μ_k^y k=1,...,n_y. The prediction error for this kriging approximation has a Gaussian distribution with variance, $\sigma_{ek}^2(\mathbf{x})$, that corresponds to the diagonal elements of the matrix

$$\mathbf{\Sigma}^y[\mathbf{P}\mathbf{\Phi}(\mathbf{x})\mathbf{P}^T + \upsilon^2\mathbf{I}]\mathbf{\Sigma}^y \tag{12}$$

where $\mathbf{\Phi}(\mathbf{x})$ is the diagonal matrix with elements $\hat{\phi}_j^2(\mathbf{x})$, j=1,...,m_c, and $\upsilon^2\mathbf{I}$ stems from error $\boldsymbol{\tau}$ [16]. An estimate for the latter is given by,

$$\upsilon^2 = \sum_{j=m_c+1}^{n_y} \lambda_j / (n_y - m_c) \tag{13}$$

corresponding to the average variance of the discarded dimensions when formulating the latent output space. The scheme for implementation of the overall metamodeling approach discussed here, as well as the risk assessment implementation discussed in the following section, is illustrated in Fig. 6.

The accuracy of the optimized surrogate model can be evaluated using a leave-one-out cross-validation approach. This approach is established by removing sequentially each of the storms from the database, using the remaining storms to predict the output for that storm, and then evaluating the error between the predicted and real responses. The validation statistics are then obtained by averaging the errors established over all storms. The most common statistics used for this purpose are the mean absolute error and the coefficient of determination which for the kth output are given by, respectively,

$$ME_k = \frac{\sum_{h=1}^{n} | y_k^h - \hat{y}_k^h |}{\sum_{h=1}^{n} | y_k^h |} \quad RD_k^2 = 1 - \frac{\sum_{h=1}^{n} \left(y_k^h - \hat{y}_k^h \right)^2}{\sum_{h=1}^{n} \left(y_k^h - \mu_k^y \right)^2} \tag{14}$$

where $\hat{y}_k^h$ denotes the approximation for the kth output for the hth hurricane in the database, an approximation established by removing that hurricane from the database when implementing the surrogate model. For the entire output vector, the representative statistics can be obtained by averaging over the individual outputs.

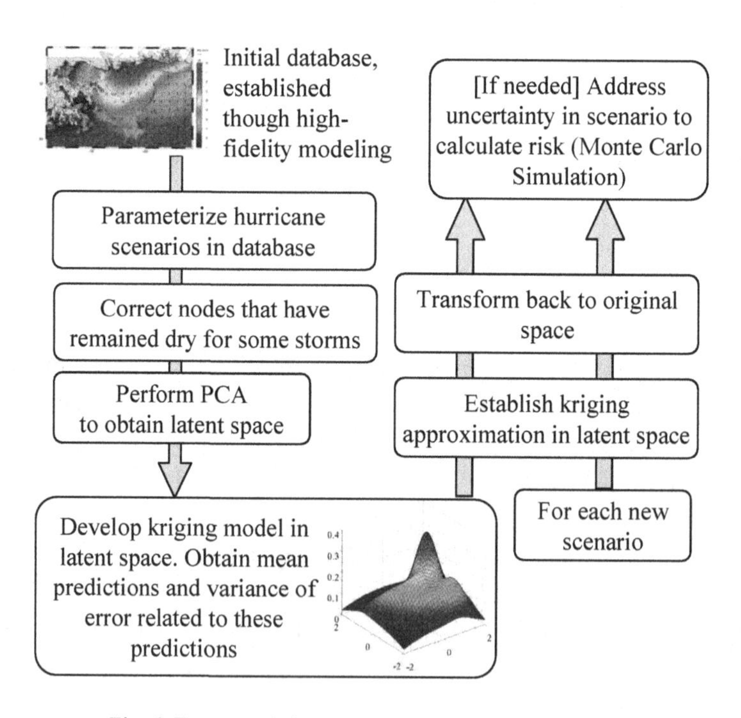

Fig. 6. Framework for surrogate model implementation.

4. Example Surrogate Modeling Applications

This section presents two case studies for surrogate modeling implementation, the first corresponding to the Hawaiian Islands and the second to the New Orleans region. In the first case study, the database was developed specifically for real-time risk assessment through surrogate modeling, whereas the second case study uses an existing database. Before proceeding with these case studies, the implementation of probabilistic risk assessment through surrogate modeling is discussed.

4.1. *Implementation for risk assessment*

Based on the surrogate model discussed in the previous section, an approximation $\hat{\mathbf{y}}(\mathbf{x})$ to the hurricane response $\mathbf{y}(\mathbf{x})$ can be obtained very efficiently for any new hurricane scenarios, and this approximation can be then adopted for estimation of the hurricane risk. For the kth dimension of the response output, this risk can be expressed in terms of $y_k(\mathbf{x})$ and assessed through its approximation $\hat{y}_k(\mathbf{x})$ while also adjusting for the existence of the prediction error, once the probability $p(\mathbf{x})$ describing the uncertainty in the hurricane model parameters is defined. For real-time risk evaluation, i.e., in cases where we are interested in predicting the risk impact of an approaching hurricane prior to landfall, $p(\mathbf{x})$ is selected through information given by the National Hurricane Center that provides the anticipated hurricane characteristics as well as standard statistical errors associated with these estimates (as a function of time to landfall). Figure 7 shows an example such a prediction.

Then risk for k^{th} dimension of the response H_k is expressed by the integral

$$H_k = \int_X h_k(\mathbf{x})p(\mathbf{x})d\mathbf{x} \tag{15}$$

where $h_k(\mathbf{x})$ is the risk occurrence measure that ultimately depends on the definition for H_k. Through appropriate selection of $h_k(\mathbf{x})$, different risk quantifications can be addressed[9]. For example, if H_k corresponds to the expected value of the response $y_k(\mathbf{x})$, we have $h_k(\mathbf{x}) = \hat{y}_k(\mathbf{x})$, whereas if it corresponds to the probability that the response $y_k(\mathbf{x})$ will exceed some threshold β_k we have

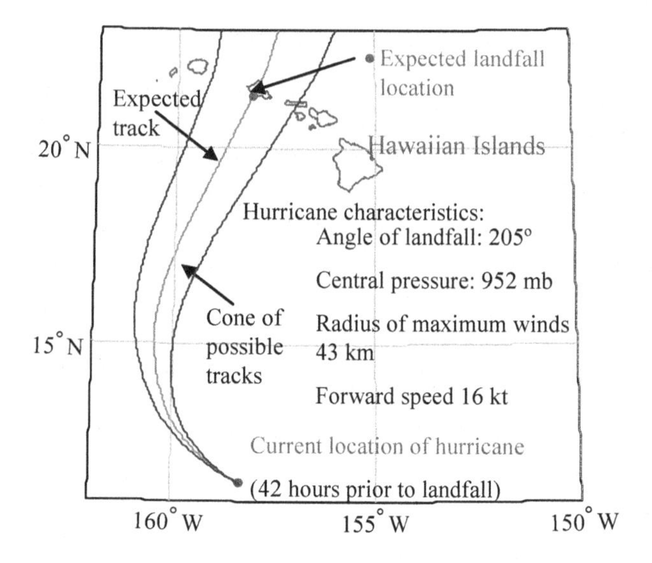

Fig. 7. Hurricane track and characteristics for the case study for the Hawaiian Islands.

$$h_k(\mathbf{x}) = \Phi\left(\frac{\hat{y}_k(\mathbf{x}) - \beta_k}{\hat{\sigma}_{ek}(\mathbf{x})}\right) \tag{16}$$

where Φ(.) denotes the standard Gaussian cumulative distribution function. If the prediction error for the metamodel is ignored [$\sigma^2_{ek}(\mathbf{x})$ is not computed and approximated to be zero], then (16) simplifies to the indicator function, equal to 1 if and zero if not.

The risk integral in (15) can be estimated by Monte Carlo simulation: using n_s samples of $\mathbf{x}$ simulated from probability density $p(\mathbf{x})$, with the jth sample denoted by $\mathbf{x}^j$, an estimate for H_k is given by

$$\hat{H}_k = \frac{1}{n_s}\sum_{j=1}^{n_s} h_k(\mathbf{x}^j) \tag{17}$$

With the adoption of the kriging metamodel, which allows for use of a large number for n_s, the risk can be then efficiently and accurately estimated with only a small computational burden.

4.2. *Hawaiian Island case study*

Since 1950, only five hurricanes have caused significant damage in Hawaii (Nina 1957, Dot 1959, Iwa 1982, Estelle 1986, and Iniki 1992). Despite the infrequency of hurricanes in Hawaii, the potential impacts are significant: severe damage to air and sea ports, island-wide power and communications outages, destruction of a large percentage of homes, and hundreds of thousands of people seeking shelter[21].

A high-fidelity database was developed in this case specifically for implementation of surrogate modeling to support real-time risk assessment for future hurricanes, through the framework discussed in Section 4.1. Since the historical record for selecting potential storms is sparse, characteristic storm parameters and tracks were selected based on historical storms and input from the National Weather Service. The database was developed by varying five hurricane parameters: landfall location, track angle, central pressure, radius of maximum winds, and forward speed. Hurricanes impacting Hawaii transit from east to west and then curve or re-curve to the north. Five tracks selected for simulation with landfall angles of 120, 150, 180, 210, and 240 deg (Fig. 4), measured clockwise from north. These tracks were shifted through the Hawaiian Islands to give 28 landfall locations (including "landfalls" between the islands). Central atmospheric pressures simulated were 940, 955, and 970 mb, roughly representing Category 2 to 4 hurricanes. Radii of maximum winds simulated were 30, 45, and 60 km, and forward speeds were 3.9, 7.7, and 11.6 kt. Not all combinations of parameters were considered, based on historical trends. Approximately 1500 storms were modeled.

Waves and storm surge were modeled for Hawaii using the high-fidelity, high-resolution numerical models described in Sec. 2. The modeling included tightly-coupled wave and surge modeling on a two-dimensional, finite-element grid. In Hawaii, because of steep offshore bottom slopes, runup can be an important component of total inundation. Runup was modeled in the original study[9,10] in a large number of transects across the coast of each island. These used a one-dimensional Bousinesq analysis forced by surge and wave information at reference points within each transect. The runup implementation is independent of the source of this information, that is, it can either stem from a high-fidelity numerical model or a surrogate model approximation, as such the approach is easily integrated within any metamodeling framework. Since the focus here is on the metamodeling aspects, this extension is not discussed further. Note that similar extensions can be established for other response quantities relying upon surge/wave information such as damage to coastal structures.

To accurately calculate waves and surge in Hawaii, the fringing reefs, wave breaking zones, and the channels in the nearshore region must be resolved. Additionally, to include wave generation as hurricanes track toward Hawaii, the mesh domain must be sufficiently large to cover the generation region to the south of Hawaii. To include both high resolution and a large domain, a finite-element approach was adopted. The computational grid extends from the equator to 35 deg north and from 139 to 169 degrees west. Bathymetry and topography used to generate the mesh were extracted from several datasets, including bathymetric and topographic lidar, shipboard multi-beam bathymetry, and National Oceanographic and Atmospheric Administration bathymetric databases. The U.S. Geological Survey National Gap Analysis Program (GAP) landcover database was used to specify Manning n bottom friction coefficients based on vegetation cover or land use type[3]. Nearshore reefs were identified by hand and given a large frictional value of 0.22, and oceanic areas with no other classification were specified to have a friction value of 0.02. To reduce the overall computational effort, two meshes were developed, both of which covered the same region, but one mesh had high resolution on Oahu and Kauai and the other had high resolution on Maui and the Big Island. Mesh resolution varied from 5000 m in deep water to 30 m in the nearshore and upland areas. The grids contain approximately 3 million elements and 1.6 million nodes as shown in Fig. 8.

Storm surge and waves were modeled with the unstructured, phase-averaged wave model SWAN[13] tightly coupled to the ADCIRC circulation model[21]. Both models were applied on the same unstructured mesh. The models were driven with wind and pressure fields generated with a planetary boundary layer model[21], applying a Garratt formulation for wind drag. SWAN was applied with 72 directional and 45 frequency bins. The standard SWAN third-generation physics were applied with Cavaleri and Malanotte-Rizzoli[23] wave growth and shallow-water triads. The ADCIRC time step was 1 s, the SWAN time step was 600 s, and two-way coupling was performed after each SWAN time step. Radiation stresses were computed by SWAN and passed to ADCIRC to force nearshore wave setup and currents, and water levels and currents were computed by ADCIRC and passed to SWAN. Simulations were run at a constant high tide of 0.4 m Mean Tide Level. Tidal variations in space and time were neglected. The inundation represented by ADCIRC is the still-water inundation, which is an average water level over approximately 10 min (averaged over individual waves). Coupled SWAN and ADCIRC simulations required approximately 2000 CPU hours on a high-performance computer for each storm. Kennedy *et al.*[10] provide model validation for water levels due to tides and for waves and water levels induced by Hurricane Iniki on Kauai.

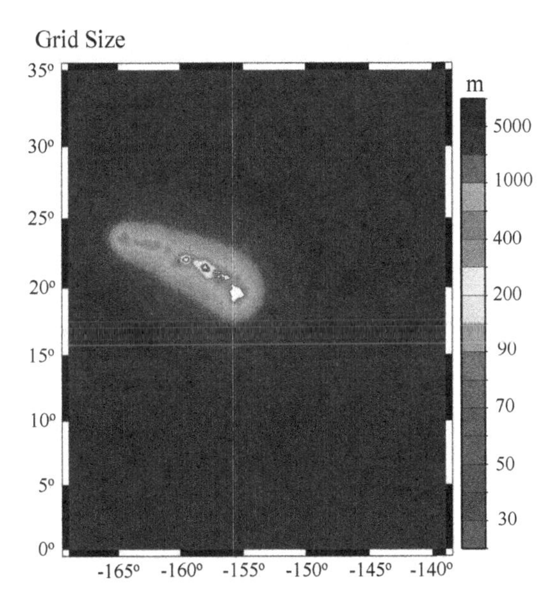

Fig. 8. Grid details for the case study for the Hawaiian Islands.

After the suite of storms was simulated, the scenarios were used to create a kriging surrogate model of storm response for the maximum surge inundation and the maximum wave height for each island separately. Here the results for the Island of Oahu are presented. The grids considered for these outputs consist of close to 10^4 points for the wave height and more than $3*10^4$ points for the surge. A single metamodel was constructed to simultaneously predict both these outputs utilizing kriging combined with PCA. For the PCA, m_c=60 latent outputs are used, which accounts for 97.1% of the total variability in the initial output and reduced the size of the matrices that need to be stored in memory by close to 90.0%. The metamodel is optimized based on (10), and its performance is validated through the leave-one-out cross-validation (discussed Sec. 3.4) for two different error statistics (i) the mean absolute error and (ii) the coefficient of determination. Surrogate model average mean errors for wave height and surge level are approximately 3.5 and 6.4%, respectively, whereas the coefficients of determination are 97% and 94%, respectively. Figure 5 showed a comparison with respect to the contours for the maximum wave height, whereas Fig. 9 shows a comparison with respect to the surge predictions. It is evident that the accuracy established through the surrogate model is very high.

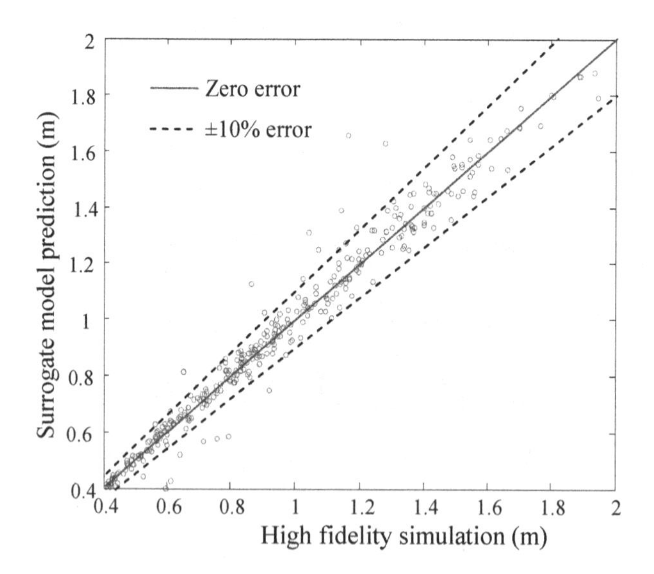

Fig. 9. Comparison of storm surge predicted by high-fidelity and surrogate models.

The computational cost for performing a single evaluation of the surrogate model is 0.1 sec on a single computational core, which corresponds to tremendous computational savings. The optimized surrogate model is, thus, able to evaluate the maximum storm response at a cost that is more than seven orders of magnitude less than the high-fidelity models with comparable accuracy. This facilitates the development of standalone risk-assessment tools that can operate on personal laptops[9]. This computational efficiency can be exploited to support a probabilistic risk assessment as discussed in Sec. 4.1. In this case the errors in forecast parameters are approximated using a zero-mean Gaussian distribution with standard deviation given as a function of time to landfall[9]. Uncertainty in the parameters decreases as the storm approaches landfall. The standard deviations used for probabilistic modeling for 72 and 12 hrs from landfall are 0.85 and 0.14 deg lon for landfall location, 30 and 5 deg for track heading at landfall, 18 and 3 mb for central pressure, 3.7 and 0.5 kt for forward speed, and 6 and 1 km for radius of maximum winds. Values between these times are estimated by linear interpolation.

A probabilistic risk assessment can then be performed to provide different risk statistics, such as the response (inundation or wave height) for a given exceedance probability. For example, the 10 percent exceedance wave height is the wave height exceeded 10 percent of the time given the storm variability and model errors. As a hurricane is far from landfall, emergency managers may focus on small exceedance values (i.e., 10 percent) to plan evacuations when there is still significant uncertainty in forecasts and apply larger exceedance values (i.e., 50 percent) or deterministic simulations when the storm is near landfall to focus on recovery. Figures 10 and 11 show an illustrative implementation for the wave height for an example scenario (landfalling hurricane 42 h prior to landfall) illustrated in Fig. 7, with mean predictions $\mathbf{x}_{mean}$=[158.17° 205° 952mb 16kt 43km] and standard deviation $\sigma_{\mathbf{x}}$=[0.28° 17.5° 10.5mb 3.5kt 2.5km]. Figure 10 shows the median wave height and the wave height with probability of exceedance 10 percent, whereas Fig. 11 shows the probability that wave height of 9 meters is going to be exceeded. The risk for all these cases is estimated by (17) using n_s=2000 samples. The total evaluation time required for this risk assessment is only 45 seconds on Intel(R) Xeon(R) CPU E5-1620 3.6GHz.

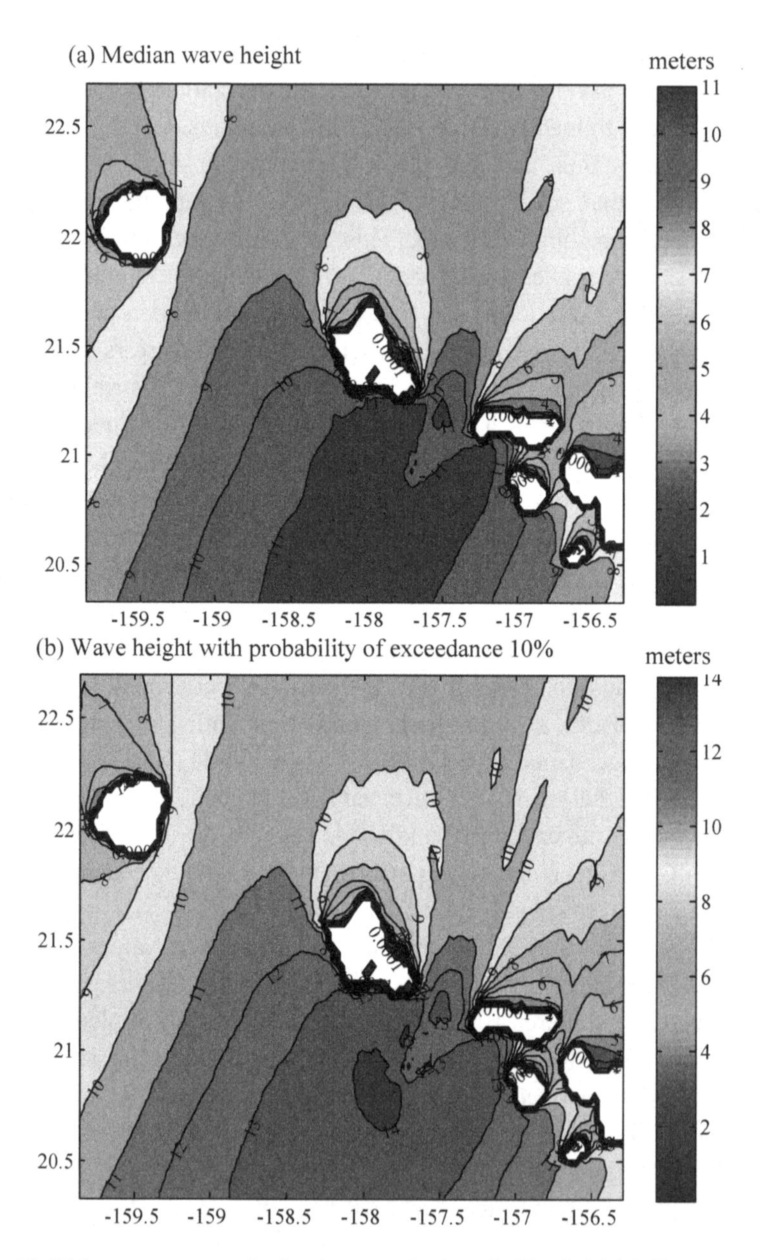

Fig. 10. Risk assessment results for the scenario show in Fig. 7. (a) Median wave height and (b) wave height with probability of exceedance 10 percent are shown.

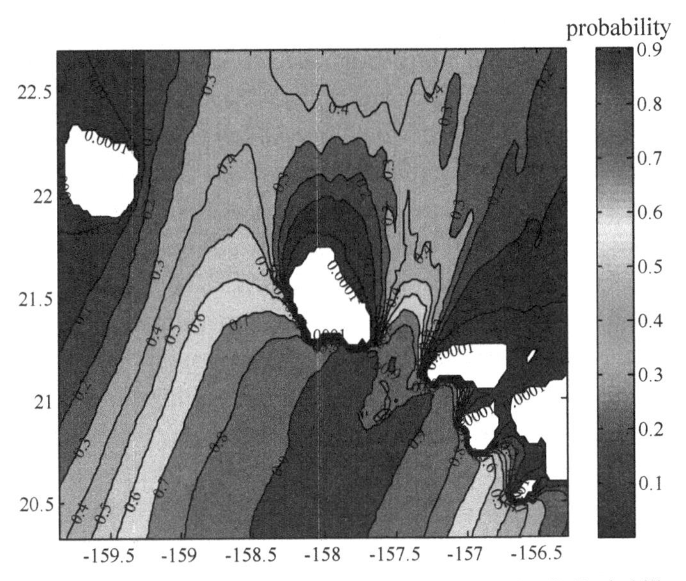

Fig. 11. Risk assessment results for the scenario show in Fig. 7. Probability of wave height exceeding 9 meters is shown.

4.3. *New Orleans, Louisiana*

The second case study considers the application of surrogate modeling using an existing database that was developed for regional flood studies in the New Orleans region[25,26]. This database consists of 304 high intensity storms and 142 low intensity storms for a total of 446 storms, and has the same hurricane parameterization as the database for the Hawaiian Islands. ADCIRC was used as the high-fidelity model for prediction of storm surge with a computational domain consisting of 2,137,978 nodes. The output considered here corresponds only to near-shore locations and only nodes that were inundated in at least 2% of the database storms are retained (at least 8 storms). This leads to a surge output of n_y=545,635 points. For developing the database, four different values for the central pressure c_p were used: 900, 930, 960 (high-intensity storms) and 975 mbar (low-intensity storms). Three different forward velocities v_f were considered, 6, 11 and 17 kt and thirteen different radius of maximum winds, R_m, 6, 8, 11, 12.5, 14.9, 17.7, 18.2, 18.4, 21, 21.8, 24.6, 25.8, and 35.6 nmi. With respect to the position characteristics, approximately 50 different landfall locations were considered corresponding to landfalls spanning from 94.5 degrees West to 88.5 degrees West. For the storm heading direction, different angles were used ranging between -60 degrees to 43 degrees (defining North = 0 degrees), based on information from regional historical storms.

A kriging metamodel combined with PCA was developed. For the PCA, m_c=50 latent outputs are used, which accounts for 97.9% of the total variability in the initial output and reduced the size of the matrices that need to be stored in memory by 88.8%. The metamodel is then optimized based on (10) and its performance is validated by calculating three different error statistics using a leave-one-out cross-validation approach. The error statistics used are (i) the mean absolute error, (ii) the coefficient of determination, and (iii) the percentage misclassification, with misclassification defined here as dry nodes predicted as wet and vice versa. When averaged over all the nodal points, these statistics are 6.64%, 0.941, and 1.19%, whereas when the focus is on nodal points that were inundated at least in 50% of the storms in the initial database, these statistics improve to 4.39%, 0.974, and 0.59%, respectively. These error estimates show that the kriging metamodel provides high-accuracy approximations to the hurricane response (small errors), with that accuracy significantly improving when looking at nodes for which higher quality information is available (nodes that were inundated in at least half of the initial storms). In particular, the high value for the coefficient of determination shows that the model can explain very well the variability in the initial database, a critical characteristic for risk assessment applications.

The optimized metamodel is used for an illustrative risk assessment. A scenario with mean value $\mathbf{x}_{mean}$=[89.9° -30° 940 14 23] [landfall longitude in degrees West, angle of landfall in degrees; central pressure in mbar; forward speed in kt; radius of maximum winds in nmi] is assumed. For defining $p(\mathbf{x})$ these parameters are assumed to follow independent Gaussian distributions with standard deviation selected as σ_x=[0.2° 1° 8 1.8 2]. The risk is estimated by (17) using n_s=2000 samples. The total evaluation time required for this risk assessment is again close to 45 seconds on Intel(R) Xeon(R) CPU E5-1620 3.6GHz. These results again correspond to a huge reduction of computational time compared to the high-fidelity model, which required a few thousand CPU hours for analyzing a single hurricane scenario. This case study, thus, further verifies that the kriging metamodel with PCA makes it possible to efficiently assess hurricane risk in real-time for a large region (high dimensional correlated outputs), providing at the same time a high-accuracy estimate for the calculated risk (small coefficient of variation since we used a large number for n_s).

Figure 12 shows an illustrative result from this risk assessment implementation, the probability $P(\zeta>\beta)$ of exceeding a specific threshold β for the storm surge ζ for two locations; location 1 has coordinates 29.8592° N, 89.8969° W (St Bernard State Park, near the Mississippi River) and location 2 has coordinates 29.9558° N, 89.9842° W (east of New Orleans and along the Mississippi River). Note that for β=0 the risk estimate $P(\zeta>\beta)$ corresponds to the probability that the location will get inundated. The risk predictions with and without accounting for the prediction error in the kriging metamodel are shown. The comparison indicates that the prediction error can have a significant impact on the calculated risk, and it will lead to more conservative estimates for rare events (with small probabilities of occurrence). This demonstrates that it is important to explicitly incorporate the local error (variance) in the risk estimation framework.

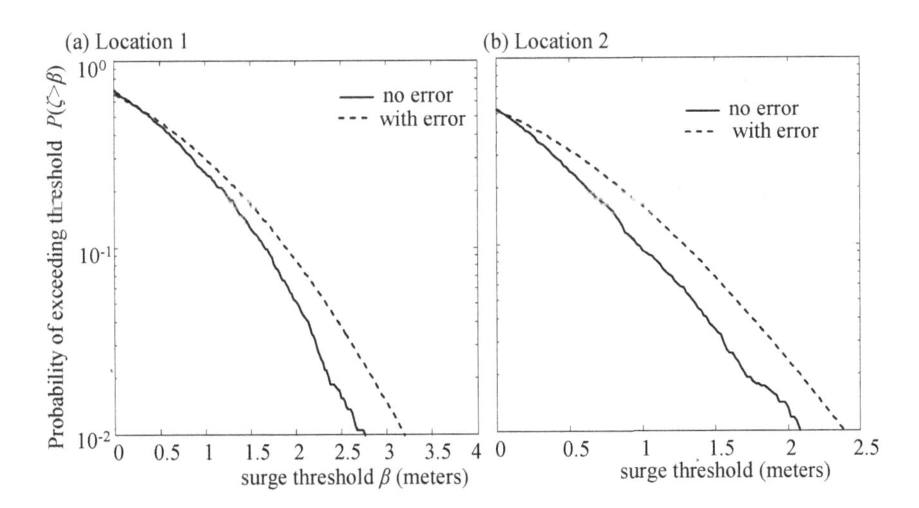

Fig. 12. Risk assessment results for the New Orleans region. Probability that surge ζ will exceed threshold β, $P(\zeta>\beta)$.

5. Summary and Future Work

High-fidelity, high-resolution wave and surge modeling is required for accurate hurricane wave and inundation simulation. Such simulations generally have a high computation cost, thus, presently limit their use for probabilistic forecasting. Surrogate modeling provides a viable tool to generate fast, accurate, and robust hurricane wave and inundation forecasts using high-fidelity simulations from an existing database. Both deterministic (for a specific event) and probabilistic estimates (considering an ensemble of probable events) can be supported in seconds to minutes on a personal computer.

Kriging is promoted in this work as a surrogate model because it is efficient and robust (requires no matrix inversions in the implementation stage), it can simultaneously provide predictions for different types of responses (e.g., inundation and waves) over an extensive coastal region, and it offers a generalized approach that is easily extendable to different regions. Principal Component Analysis (PCA) is also considered to reduce the dimensionality of the problem by projecting the initial database into a lower dimensional (latent) space using eigenvectors. This lower dimensional space is selected so that it captures a large portion of the variability observed in the initial database. This approach ultimately facilitates a significant reduction in the memory requirements for implementation of the surrogate model, e.g., in the cases considered here, a reduction of almost 90%.

The overall methodology for the surrogate modeling is first to establish a database of high-fidelity simulations that covers the ranges of storms to be forecast. Hurricanes in Hawaii and southeastern Louisiana are given as examples, and these hurricanes were parameterized based on five characteristic (landfall location, angle of approach, central pressure, forward speed, and radius of maximum winds). The next step is to correct nodes that remain dry in some of the storms, if surge for such nodes needs to be predicted. Then PCA is performed to obtain the latent space, and the kriging metamodel is finally developed in the latent space. To simulate a scenario, the hurricane parameters at landfall are entered, and the kriging approximation is immediately obtained. Then, the solution is transformed from the latent space back to the original physical variables (inundation and waves) to provide the required predictions. Because the solutions are so fast (~ 0.1 sec per scenario), Monte Carlo simulations can be quickly performed to provide probabilistic assessments within a few minutes.

Solutions are robust because quality control is done prior to construction of the database, and the kriging approximation can be optimized to offer high accuracy. The mean absolute error for the prediction for the Hawaii case study was 3.5% for wave height and 6.4% for surge. For the New Orleans case study, this error was 6.6% for the surge when considering the predictions over the entire coastal region of interest, whereas it improved to 4.4% when considering only the locations that were inundated at least 50% of the simulations in the initial database (before any corrections were applied).

Areas of future improvement to the surrogate modeling methodology include extension to extratropical storms, addition of rainfall and runoff induced flooding, integration of structural damage prediction, and improvements of the correction process for the surge (and subsequently in the accuracy of predictions) for any locations that have remained dry in some of the storms in the database of high-fidelity simulations.

Acknowledgments

This work was supported by the Storm Vulnerability Protection work unit of the Flood and Coastal Storm Damage Reduction Program, Engineering Research and Development Center, US Army Corps of Engineers. The Program Manager is Dr. Cary Talbot and the Technical Director is Mr. William Curtis. Permission to publish this work was granted by the Chief of Engineers, U.S. Army Corps of Engineers.

References

1. E. S. Gibney and E. Blake, "The Deadliest, Costliest, and Most Intense United States Tropical Cyclones 1851-2010 (and other frequently requested hurricane Facts)," NOAA Technical Memorandum NWS NHC-6, National Weather Service, National Hurricane Center, Miami, FL, (2011).
2. National Oceanic and Atmospheric Administration, "Adv. Hurricane/Post-Tropical Cyclone Sandy, October 22-29, Service Assessment," 2013.
3. S. Bunya, J. J. Westerink, J. C. Dietrich, H. J. Westerink, L. G. Westerink, J. Atkinson, B. Ebersole, J. M. Smith, D. Resio, R. Jensen, M. A. Cialone, R. Luettich, C. Dawson, H. J. Roberts and J. Ratcliff, "A High-Resolution Coupled Riverine Flow, Tide, Wind, Wind Wave and Storm Surge Model for Southern Louisiana and Mississippi: Part I—Model Development and Validation," *Monthly Weather Review,* vol. 138, pp. 345-377, (2010).

4. J. C. Dietrich, S. Bunya, J. J. Westerink, B. A. Ebersole, J. M. Smith, J. H. Atkinson, R. Jensen, D. T. Resio, R. A. Luettich, C. Dawson, V. J. Cardone, A. T. Cox, M. D. Powell, H. J. Westerink and H. J. Roberts, "A High-Resolution Coupled Riverine Flow, Tide, Wind, Wind Wave and Storm Surge Model for Southern Louisiana and Mississippi: Part II—Synoptic Description and Analysis of Hurricanes Katrina and Rita," *Monthly Weather Review,* vol. 138, pp. 378-404, (2010).

5. J. C. Dietrich, J. J. Westerink, A. B. Kennedy, J. M. Smith, R. Jensen, M. Zijlema, L. H. Holthuijsen, C. Dawson, R. A. Luettich, M. D. Powell, V. J. Cardone, A. T. Cox, G. W. Stone, H. Pourtaheri, M. E. Hope, S. Tanaka, L. G. Westerink, H. J. Westerink and Z. Cobell, "Hurricane Gustav (2008) Waves, Storm Surge and Currents: Hindcast and Synoptic Analysis in Southern Louisiana," *Monthly Weather Review,* vol. 139, pp. 2488-2522, (2011).

6. A. B. Kennedy, U. Gravois and B. Zachry, "Observations of landfalling wave spectra during Hurricane Ike," *Journal of Waterway, Port, Coastal and Ocean Engineering,* vol. 137, no. 3, pp. 142-145, (2011).

7. M. E. Hope, J. J. Westerink, A. B. Kennedy, P. C. Kerr, J. C. Dietrich, C. Dawson, C. Bender, J. M. Smith, R. E. Jensen, M. Zijlema, L. H. Holthuijsen and R. A. Luettich, "Hindcast and Validation of Hurricane Ike (2008) Waves, Forerunner, and Storm Surge," *J. Geophysical Research,* vol. 118, no. 9, pp. 4424-4460, (2013).

8. J. L. Irish, D. T. Resio and M. A. Cialone, "A surge response function approach to coastal hazard assessment. Part 2: Quantification of spatial attributes of response functions," *Natural Hazards,* vol. 51, no. 1, pp. 183-205, (2009).

9. A. Taflanidis, A. Kennedy, J. Westerink, J. Smith, K. Cheung, M. Hope and S. Tanaka, "Rapid Assessment of Wave and Surge Risk during Landfalling Hurricanes: Probabilistic Approach," *Journal of Waterway, Port, Coastal, and Ocean Engineering,* vol. 139, no. 3, pp. 171-182, (2013).

10. A. B. Kennedy, J. J. Westerink, J. M. Smith, A. A. Taflanidis, M. Hope, M. Hartman, S. Tanaka, H. Westerink, K. F. Cheung, T. Smith, M. Hamman, M. Minamide, A. Ota and C. Dawson, "Tropical Cyclone Inundation Potential on the Hawaiin Islands of Oahu and Kauai," *Ocean Modeling,* Vols. 52-53, pp. 54-68, (2012).

11. D. T. Resio, J. L. Irish and M. A. Cialone, "A surge response function approach to coastal hazard assessment – part 1: basic concepts," *Natural Hazards,* vol. 51, no. 1, pp. 163-182, (2009).

12. P. C. Kerr, A. S. Donahue, J. J. Westerink, R. A. Luettich, L. Y. Zheng, R. H. Weisberg, Y. Huang, H. V. Wang, Y. Teng, D. Forrest, Y. J. Zhang, A. Roland, A. T. Haase, A. Kramer, A. Taylor, R. R. Rhome, J. Feyen, R. P. Signell, J. Hanson, M. E. Hope, R. Estes, R. Dominguez, R. Dunbar, L. Semeraro, H. J. Westerink, A. Kennedy, J. M. Smith, M. D. Powell, V. J. Cardone and A. T. Cox, "U.S. IOOS Coastal & Ocean Modeling Testbed: Inter-Model Evaluation of Tides, Waves, and Hurricane Surge in the Gulf of Mexico," *J. Geophysical Research-Oceans,* vol. 118, pp. 5129-5172, (2013a).

13. M. Zijlema, "Computation of wind-wave spectra in coastal waters with SWAN on unstructured grids," *Coastal Engineering,* vol. 57, pp. 267-277, (2010).

14. P. C. Kerr, R. C. Martyr, A. S. Donahue, M. E. Hope, J. J. Westerink, R. A. Luettich, A. Kennedy, J. C. Dietrich, C. Dawson and H. J. Westerink, "U.S. IOOS Coastal and Ocean Modeling Testbed: Evaluation of Tide, Wave, and Hurricane Surge Response Sensitivities to Grid Resolution and Friction in the Gulf of Mexico," *J. Geophysical Research-Oceans,* vol. 118, pp. 4633-4661, (2013b).
15. A. A. Taflanidis, G. Jia, A. B. Kennedy and J. M. Smith, "Implementation/ optimization of moving least squares response surfaces for approximation of hurricane/storm surge and wave responses," *Natural Hazards,* vol. 66, no. 2, pp. 955-983, (2012).
16. G. Jia and A. A. Taflanidis, "Kriging metamodeling for approximation of high-dimensional wave and surge responses in real-time storm/hurricane risk assessment," *Computer Methods in Applied Mechanics and Engineering,,* Vols. 261-262, pp. 24-38, (2013).
17. M. Tipping and C. Bishop, "Probabilistic Principal Component Analysis," *Journal of the Royal Statistical Society: Series B (Statistical Methodology),* vol. 61, no. 3, pp. 611-622, (1999).
18. J. Sacks, W. Welch, T. Mitchell and H. Wynn, "Design and analysis of computer experiments," *Statistical Science,* vol. 4, no. 4, pp. 409-435, (1989).
19. S. N. Lophaven, H. B. Nielsen and J. Sondergaard, "Aspects of the MATLAB toolbox DACE, in: Informatics and mathematical modelling report IMM-REP-2002-13," Technical University of Denmark, (2002).
20. K. F. Gilbert, Interviewee, [Interview]. (September 2010).
21. R. A. Luettich and J. J. Westerink, "Formulation and numerical implementation of the 2D/3D ADCIRC finite element model.," (2004). [Online]. Available: http://adcirc.org/adcirc_theory_2004_12_08.pdf.
22. E. F. Thompson and V. Cardone, "Practical modeling of hurricane surface wind fields," *Journal of Waterway, Port, Coastal and Ocean Engineering,* vol. 122, no. 4, pp. 195-205, (1996).
23. L. Cavaleri and P. Malanotte-Rizzoli, "Wind wave prediction in shallow water: theory and applications," *J. Geophysical Research,* vol. C11, pp. 10961-10973, (1981).
24. Z. Demirbilek and O. G. Nwogu, "Boussinesq Modeling of Wave Propagation and Runup Over Fringing Coral Reefs, Model Evaluation Report, ERDC/CHL TR-07-12," U.S. Army Engineer Research and Development Center, Vicksburg, Mississippi, (2007).
25. U.S. Army Corps of Engineers, "Louisiana Coastal Protection and Restoration (LACPR) – Final Technical Report and Comment Addendum," USACE, Washington, D.C., (2009a).
26. U.S. Army Corps of Engineerings, "Interagency Performance Evaluation Task Force, Final Report on "Performance Evaluation of the New Orleans and Southeast Louisiana Hurricane Protection System," Volume VIII: Engineering and Operational Risk and Reliability Analysis," U.S. Army Corps of Engineerings, Washington, D.C., (2009b).

CHAPTER 8

STATISTICAL METHODS FOR RISK ASSESSMENT OF HARBOR AND COASTAL STRUCTURES

Sebastián Solari[1] and Miguel A. Losada[2]

[1] *IMFIA-FING, Universidad de la República, Julio Herrera y Reissig 565, Montevideo, 11300, Uruguay*
ssolari@fing.edu.uy; ssolari@ugr.es
[2] *IISTA, Universidad de Granada, Av. del Mediterráneo s/n Edif. CEAMA, Granada, 18071, Spain*
mlosada@ugr.es

Resumé

Design standards and recommendations require designers to analyze the performance of structures from a global perspective, calculating not only their reliability but also their serviceability and operationality during their entire useful life. This is achieved through the use of ultimate limit states (ULS), serviceability limit states (SLS) and operational limit states (OLS) (e.g.: ROM 0.0, 2001, ROM 1.0, 2009). These states cover the entire range of climate conditions: minima or calms, central or normal and maximum or severe conditions (central conditions are the range of values the variable normally takes and are those of relatively high probability, i.e., the bulk of the data or the mean climate). Given that the joint probability distributions entail decisions that can produce effects at all levels, the analysis (selection, estimation and handle) of such statistical distributions and the simulation of time series of the met-ocean variables are a relevant part of the analysis and evaluation of the Harbor or Coastal Investment Project. This chapter is devoted to present new methods and tools for helping the engineers to design and analyze alternatives of harbor and coastal structures under the frame of risk assessment and decision makers.

1. Introduction: Justification and Motivation

A harbor is designed to facilitate the port and logistic operations associated with maritime transportation and its relation to other transportation modes, and the integral management of vessels. Among other things, it has infrastructures that are directly related to the safety, use, and exploitation of vessels, sea oscillation control, land use and exploitation of the surrounding area, and vehicle access and transit (road and railroad traffic).

Similarly, an Integrated Coastal Zone Management aims to facilitate the orderly and sustainable use and exploitation of the coastal environment. This includes inter alia the correction, protection and defense of the shoreline, the generation, conservation and nourishment of beaches and bathing areas, as well as the interchange of transversal land-sea flows of various substances.

The design of the maritime structure is based on its performance and its interaction in plan as well as in elevation with the agents envisaged in the project design. These agents can be related to gravity, the environment, soil characteristics, use and exploitation, materials, and construction procedures. Through the analysis of structure´s performance, it is possible to describe and classify the mechanisms that lead to the failures or operational stoppages of the structure (failure or stoppage modes). Their occurrence can produce economic, social and environmental consequences.

One of the tasks of the project is to verify the fulfillment of the project requirements and to determine when, why, under which conditions, and how often can those modes occur. In general, the verification follows the procedure known as the "limit states method" (see ROM 0.0, 2001), which entails verifying failure or stoppage modes solely for those project design states which have foreseeable limit situations. These limit situations can arise in the structure because of problems related to its: (1) structural resistance (safety, ultimate limit states, ULS), (2) structural and formal properties (serviceability limit states, SLS), and (3) use and exploitation (operational limit states, OLS).

The Spanish Recommendations for Maritime and Harbor Structures, ROM 0.0 (2001), require that the design of any maritime structure should satisfy fixed safety, serviceability, and use and exploitation requirements.

They are specified by enclosing the probability of exceedance of the modes of failure binding the safety and serviceability of the structure during its useful life, and the probability of no-exceedance of the modes of stoppage, related with the operationality during the year.

When the predominant agents of the maritime structure are climate-related, like for breakwaters, it is advisable to temporally sequence of the useful life of the structure by using the sequence of states curves of some agent descriptors like the significant wave height. These curves should help to identify the safety and serviceability threshold values of the agents whose exceedance can significantly affect the performance of the structure. Moreover, it is possible to identify the threshold values of the use and exploitation agents, whose exceedance can significantly affect the operationality of the structure.

Depending on the design requirements (safety, serviceability or operationality), the distribution functions can belong to the extreme upper or lower values, or fall within the middle group of values that the variable can take during the time period. Generally speaking, extreme upper or lower values correspond to the occurrence of extreme events, included in extreme work and operating conditions (extreme regime), whereas the middle values refer to normal work and operating conditions (middle regime).

The evaluation of the Investment Project must tackle and effectively deal with the optimization of the economic and financial benefits during the useful life of the structure, and the economic effects of the different operations and agents involved. It is the job of the project engineers, whether in the public or private sector, to provide the necessary information to define and optimize the objective function that evaluates the economic, social, and environmental importance of the structure and the consequences of its failure.

Given that the joint probability distributions entail decisions that can produce effects at all levels, the analysis (selection, estimation and handle) of such statistical distributions are a relevant part of the analysis and evaluation of the Harbor Investment Project. The final goal of the project is to assess the risk of the system (harbor or coastal zone) because of the maritime structure. Here risk is defined as the probability of occurrence of the modes of failure and stoppage of the maritime structure times their consequences.

As a general rule, this analysis should be carried out within the framework of Decision Theory. This naturally signifies using optimization techniques for an objective function, subject to restrictions. Examples of such restrictions are the satisfaction of (1) the joint probability values recommended in ROM 0.0, 2001; (2) the economic and financial profitability of the Investment Project and its social consequences; and (3) the environmental requirements specified in the corresponding environmental analysis, and (in Europe) in the Water Frame and Flooding Directives, and, after March 2015, a new directive related with the Integrated Coastal Zone Management.

The application of this design loop requires that the uncertainty of the verification equations of the failure and operational stoppages modes is equity, and a fair contribution of each of them to the global probability of failure and operational stoppage. Furthermore, for maritime works and breakwaters, the result of the analysis depends on the specific sequence of sea states (storms and calms) that have arisen during the structure's useful life. If the useful life were "repeated", the result would be different each repetition, given the random nature of the occurrence of the agents and their manifestations.

During the useful life of the structure only a specific sequence of sea states will occur. It is almost impossible to know, a priori, which one will be. The best technique to work out this problem is to repeat (to simulate) the useful life of the structure many times. On the basis of the results obtained, to elaborate a statistically significant sample, which would make it possible to infer the failure probability of each of the principal modes and the joint probability of all the modes of failure and stoppage.

The method most widely used to simulate sequences of storms and calms involves developing joint or conditioned distributions for the random variables of storm occurrence, intensity and duration, using Poisson distribution and a generalized Pareto distribution for the first two and binding intensity and duration (ROM 0.0, 2001; Payo et al. 2008). Solari and Losada (2011) improved this approach incorporating the non-stationary (seasonal and inter-annual) behavior of the met-ocean variables in the simulation and modelling their time dependence.

In the following sections we summarized the most recent advances to assess the risk of a maritime structure. Specifically, we try to answer the following questions related with the description and simulation of the agents (meteorological and oceanographical variables: wind speed, wave height, etc) and of the actions (dynamic and kinematic variables: pressures, velocities, etc.):

(a) Which is the probability (marginal) distribution of the variable? Is it possible to obtain a good fit with a parametric function?
(b) Does such distribution fit the entire range of values of the variables, bulk and tail data?
(c) Is there a procedure to choose the threshold value of the upper tail domain? In this case, which is its distribution function, what are the values of the high-return period quantiles, and what is their uncertainty?
(d) Is it possible to analyze the seasonal and inter-annual variability of the variables? How can they be incorporated in the probability distribution functions?
(e) In the frame of risk assessment, which are the most adequate methods to simulate multidimensional time series of met-ocean variables?

2. Unified Distribution Models for Met-ocean Variables

There is a need of probability models for met-ocean variables that quantify as accurately as possible their frequency of occurrence and their uncertainty, as well as its seasonal and inter-annual variability. Such models would preferably cover the entire range of values of the variables and would thus model both the central distribution and the tails. This last aspect is particularly important when the serviceability and the use and exploitation depend not only on storm conditions (maximum regime) but also on central and calm conditions (central and minimum regimes, respectively), as for the integrated coastal zone management or the design and maintenance of navigation channel.

This section is devoted to answer some of the above questions. Firstly, the mixture distribution models are analyzed and their capability to work out with all the population, bulk and tail, data. Special attention is given to the very common cases of bi- and multi-modal random variables and of circular variables. Next, the fit of distribution of the extremes of a single random variable is explored. Finally, these models are broaden to non-stationary conditions (for different scales of variability).

2.1. *Stationary Mixture Distribution Models*

2.1.1. *Central-GPD distribution*

The traditional approach for modelling the probability distribution of a met-ocean variable is to differentiate the study of the central regime (i.e., the bulk of the data) from the study of the maximum and minimum regimes. For the central regime, either certain standard distributions are tested or an empirical distribution is used. These approximations necessarily limit the validity of the results to the central regime because standard distributions do not usually provide a good fit for the tails (Ochi, 1998), whereas extrapolation is not possible with an empirical distribution.

Following Vaz De Melo Mendes and Freitas Lopes (2004), Behrens et al. (2004), Tancredi et al. (2006), Cai et al. (2008) and Furrer and Katz (2008), Solari and Losada (2012a, 2012b) proposed a parametric mixture model that differentiates the three populations: (i) a central population for the mean regime; (ii) a minima regime population for the lower tail; (iii) a maxima regime population for the upper tail. The thresholds that define the limit between populations are parameters of the model, and are thus estimated in the same way as the other parameters. The proposed model, given in (1), consists of a truncated central distribution function for the bulk of the data and two generalized Pareto distributions (GPD) for the tails.

$$f(x) = \begin{cases} f_m(x)F_c(u_1) & x < u_1 \\ f_c(x) & u_1 \leq x \leq u_2 \\ f_M(x)(1 - F_c(u_2)) & x > u_2 \end{cases} \tag{1}$$

where $f_c(x)$ is the density function selected for the central part, $f_m(x)$ is the lower tail GPD and $f_M(x)$ is the upper tail GPD; u_1 and u_2 are lower and upper thresholds where the central distribution is truncated; $F_c(u_1)$ and $\left(1 - F_c(u_2)\right)$ are scale constants for the lower and upper GPD respectively.

Upper and lower tail GPD are given by (2) and (3) respectively (e.g. Coles, 2001):

$$f_m(x|x < u_1) = \frac{1}{\sigma_1}\left(1 - \frac{\xi_1}{\sigma_1}(x - u_1)\right)^{-\frac{1}{\xi_1}-1} \quad (2)$$

$$f_M(x|x > u_2) = \frac{1}{\sigma_2}\left(1 + \frac{\xi_2}{\sigma_2}(x - u_2)\right)^{-\frac{1}{\xi_2}-1} \quad (3)$$

where $\xi_1, \xi_2 \neq 0$ are the shape parameters (for $\xi_1, \xi_2 = 0$ GPD reduces to an exponential distribution), $\sigma_1, \sigma_2 > 0$ are the scale parameters and u_1 and u_2 are the position parameters (lower and upper thresholds respectively). For minima GPD (f_m) is $u_1 + \sigma_1/\xi_1 \leq x \leq u_1$ if $\xi_1 > 0$, and $x \leq u_1$ if $\xi_1 > 0$. Conversely, for maxima GPD (f_M) is $u_2 \leq x \leq u_2 - \sigma_2/\xi_2$ if $\xi_2 < 0$, and $x \geq u_2$ if $\xi_2 > 0$.

For the density function (1) to be continuous and to have lower bound equal to zero, the following relations must be fulfilled

$$\sigma_1 = -\xi_1 u_1 \;\; ; \;\; \xi_1 = -\frac{F_c(u_1)}{u_1 f_c(u_1)} \;\; ; \;\; \sigma_2 = \frac{1 - F_c(u_2)}{f_c(u_2)} \quad (4)$$

In several mid-latitude location the Log-Normal (LN) and the Weibull biparametric (WB) distributions had been found adequate for the central distribution when modelling significant wave heights ($f_c = f_{LN}$) and mean wind speeds ($f_c = f_{WB}$) respectively (Mendonça et al., 2012; Solari and Losada, 2012a; Solari and van Gelder, 2012).

The LN distribution has position parameter μ_{LN} and scale parameter $\sigma_{LN} > 0$; the WB distribution has scale parameter $\alpha_{WB} > 0$ and shape parameter $\beta_{WB} > 0$. In both cases, using relations (4), the resulting mixture model has five parameters: the two parameters of the central distribution, the two thresholds u_1 and u_2 and the upper tail shape parameter ξ_2. For details on the procedure for fitting the parameters of the distribution the reader is referred to Appendix A and to Solari and Losada (2012a).

2.1.2. *Bi- and multi-modal distributions*

Some met-ocean variables have bi- or multi-modal probability distributions. In these cases it is possible to use mixture distributions (5) to model the probability distribution of the variables.

$$f(x) = \sum_{i=1}^{N} \alpha_i f_i(x) \text{ with } 0 \leq \alpha_i \leq 1; \ \sum_{i=1}^{N} \alpha_i = 1 \qquad (5)$$

Distributions $f_i(x)$ may be from a single family or from multiple families and, if it is requiered to adequately represent the tails of the distribution, mixture distribution (1) presented on §2.1.1. can be used for $f_i(x)$ as well.

A variable that typically shows bimodal distribution is the peak wave period, particularly in areas where there is a clear distinction between sea and swell. In such cases a mixture distribution made of two log-normal distributions (6) has proven to be adequate (see e.g Solari and van Gelder, 2012)

$$f(x) = \alpha f_{LN_1}(x) + (1 - \alpha) f_{LN_2}(x) \text{ with } 0 \leq \alpha \leq 1 \qquad (6)$$

2.1.3. *Circular variables*

Two other variables that typically show multi-modal distributions are wave and wind directions. In these cases the circular behavior of the variables should be considered, using circular or wrapped distribution functions for constructing the mixture distribution

$$f_w(x) = \sum_{i=1}^{N} \alpha_i f_{w,i}(x) \text{ with } 0 \leq \alpha_i \leq 1; \ \sum_{i=1}^{N} \alpha_i = 1; \ 0 \leq x \leq 2\pi \qquad (7)$$

Although in principle any linear distribution $f(x)$ can be wrapped by means of $f_w(x) = \sum_{i=0}^{+\infty} f(x \pm 2k\pi)$, being $0 \leq x \leq 2\pi$, there are some distributions already defined for circular variables, as von Misses or Wrapped Normal distributions (Fisher, 1993), the latter one given by,

$$f_{WN}(x) = \frac{1}{2\pi}\left(1 + 2\sum_{p=1}^{\infty} \rho^2 \cos p(x - \mu)\right) \quad (8)$$

where parameters μ and ρ represent directional mean and dispersion, with $0 \leq x \leq 2\pi$.

2.2. *The Use of Mixture Models for Estimating High Return Period Quantiles*

Parametric modeling for extreme conditions of met-oceanic variables is required when attempting to infer unrecorded conditions from available data.

Extreme value theory states that the distribution of the maxima or minima of an independent and identically distributed (i.i.d.) series of n elements tends to have one of the three forms of the generalized extreme value distribution. It also states that the distribution of the values that exceed a given threshold of a series of i.i.d. data tends to have a GPD when the threshold tends toward the upper bound of the variable (see e.g. Castillo 1987, Coles 2001, Kottegoda and Rosso 2008).

These results establish the theoretical foundation for the two most widely accepted methods for modeling the extremes of several geophysical variables: the block maxima method (usually annual maxima) and the over-threshold method (Leadbetter, 1991). When the entire time series of data is available, the use of the block maxima method is associated with a significant loss of information concerning extreme events. The over-threshold method, on the other hand, uses the data more efficiently by considering more than one sample per year. In this sense, the over-threshold method is preferred over the block maxima method when an entire time series is available (Madsen et al., 1997).

When applying the over-threshold approach on an i.i.d. series it is straightforward to obtain high return period quantiles from the marginal distribution of the upper tail of the distribution. To this end the upper GPD provided by the mixture model (1) presented on §2.1.1 can be used.

On the other hand, when the series shows a tendency to form clusters (e.g. storm events), as is the case of most met-ocean variables, the distribution of the extreme values would depends not only on the marginal distribution of the upper tail (as is the case with i.i.d. series) but also on the characteristics of the clusters.

In this latter case there are two ways of addressing the problem of the declustering of extreme values: (1) by declustering the data, constructing an i.i.d. series, or (2) by accounting for the dependence of the original series (i.e. analyzing the clustering of the extreme values).

The first framework is the most widely adopted (Coles, 2001) and includes the POT method that is discussed in §2.2.1. In this case it is no feasible to directly use the mixture model presented on §2.1.1 for estimating extreme values, however it provides an automatic and objective way to estimate the threshold required for applying the POT method, as described next.

In the second framework it is feasible to use directly the mixture model presented on §2.1.1. However, more sophisticated statistical models are required for the estimation of the cluster characteristics, and particularly for accounting for cluster duration. The interested reader is referred to Eastoe and Tawn (2012), Fawcett and Walshaw (2012), and reference therein.

2.2.1. *Using cluster maxima: the POT method*

The POT method is considered to be the declustering method most commonly used by coastal engineers. References can be found in Davison and Smith (1990), although, as pointed out in the discussion of the paper, the idea is older. Given a chosen threshold, the exceedance values that are separated by less than a given minimum time span are assumed to form a cluster. The main assumption is that each cluster is generated by the same extreme event; then, the maximum record value is taken. This leads to the construction of a POT series of independent observations.

Once a POT series is constructed a distribution is fitted in order to estimate high return period quantiles. The GPD is the distribution most commonly used for fitting POT data since it use is supported by mathematical considerations (Pickands, 1975), although some authors argue that the use of other distribution is equally valid (e.g. Mazas and Hamm, 2011). Nevertheless, it is clear that the threshold is an essential parameter for applying the POT method, whether or not the GPD is used for fitting the data. However, despite the importance of the threshold in the analysis of extreme events, the existing methods for threshold identification are, in some way, based on subjective judgment.

Two common ways of choosing a threshold are based on expert judgment. One way is to select a fixed quantile corresponding to a high nonexceedance probability, usually 95%, 99%, or 99.5% (see, e.g., Luceño et al. 2006 or Smith 1987). The other way is to impose a minimum value to the mean number of clusters per year.

There are other methods that provide some guidance for threshold identification and limit the subjectivity of its selection: the graphical methods (GM), Coles 2001, the optimal bias-robust estimation (OBRE) method, Dupuis 1998, the methods proposed by Thompson et al. (2009) and by Mazas and Hamm (2011), both based on the stability of the parameters of the GPD, and the method proposed by Solari and Losada (2012a, 2012b) based on the use of the mixture models. These methods are discussed below; however the reader should be aware that this is not a comprehensive discussion of all the existing methods for threshold selection, and that there are some methodologies that will not be analysed (see e.g. Rosbjerg et al., 1992).

Graphical methods, Coles 2001

The Mean Residual Life Plot (MRLP) designates the graph of the series $\{u, 1/n_u \sum_{i=1}^{n_u}(x_i - u)\}$, where $\{x_i\}$ is the series of data such that $u < x_i < x_{max}$ and n_u is the number of elements of $\{x_i\}$. The MRLP should be linear above the threshold u_0 from which point the GPD provides a good approximation of the distribution of the data. Similarly, the estimates of ξ, and $\sigma^* = \sigma - \xi_u$, where ξ and σ are the shape and scale parameters of the GPD, should be constant above the threshold u_0. The graphical method consists of creating such plots, based on which threshold u_0 is then visually estimated (for more details see Coles, 2001).

A major advantage of the GM is that its implementation is straightforward and it reduces the subjectivity associated with threshold selection. However, the GM has a remaining subjective component that requires human judgment and cannot be automated, and it is unable to provide the uncertainty of the threshold estimation.

OBRE method, Dupuis 1998

Dupuis (1998) proposed performing parameter estimation and selecting the threshold by introducing the optimal bias-robust estimator (OBRE), which is an M estimator (see de Zea Bermudez and Kotz, 2010) that attributes weights (equal to or less than one) to the data used in parameter estimations. Dupuis (1998) suggests using these weights as a guide for choosing the threshold.

Method of the parameter σ^*, Thompson et al. 2009

The method proposed by Thompson et al. (2009) consists of calculating parameter σ^* for a series of thresholds $\{u_i\}$. Above u_0 afterwhich the GPD provides a good approximation to the data, the series $\{\sigma^*_{u_i} - \sigma^*_{u_{i-1}} | u_i > u_0\}$ should have a zero-mean normal distribution. After the application of the chi-square normality test to the series $\{\sigma^*_{u_i} - \sigma^*_{u_{i-1}}\}$, u_0 is selected as the first value of the series $\{u_i\}$ for which the hypothesis test does not reject the null hypothesis (for more details see Thompson et al., 2009). The method requires the previous definition of four parameters: the level of significance used for the hypothesis test and the threshold series $\{u_i\}$, defined by a minimum threshold, a maximum threshold, and the number of intermediate thresholds. Thompson et al. (2009) report that, in the resolution of their problem, they obtain good results using a series of 100 thresholds between the quantile corresponding to 50% of the sample and the minimum between the quantile corresponding to 98% and the value that is exceeded only by 100 data.

Method of parameter's stability, Mazas and Hamm (2011)

Mazas and Hamm (2011) recommend the use of the graphical method, based on the stability of the parameters ξ and σ^*, in conjunction with an analysis of the mean number of peaks obtained per year. These authors recommend selecting the lowest threshold of the highest domain of stability of ξ and σ^*, while checking that the mean number of peaks per year is roughly between 2 and 5.

Mixture models with assimilated thresholds, Solari and Losada (2012a,b)

Solari and Losada (2012a,b) proposed to use the upper threshold u_2 estimated with mixture distribution model (1) to construct the POT series. To this end, the minimum time between the end of an exceeding event and the start of a new one can be estimated in such a way that the resulting POT series meets the Poisson hypothesis (see Cunnane, 1979) and does not show lag one autocorrelation (see Solari and Losada 2012b for details). This threshold estimation methodology is automatable, requiring only the definition of the central distribution $f_c(x)$ in model (1) that results more adequate for the bulk of the data.

2.3. *Non-Stationary Mixture Distribution Models*

The state curves of the met-ocean variables and other geophysical variables present seasonal (annual cycle) and inter-annual (i.e., long-term cycles of over a year) cycles. A simple way to incorporate this behavior is to describe the parameters of the probability distributions by Fourier series with the appropriated mean time period, i.e. the year

$$\theta(t) = \theta_0 + \sum_{k=1}^{N_k} (\theta_{ak} \cos 2k\pi t + \theta_{bk} \sin 2k\pi t) \tag{9}$$

where t is the time measured in years (see, e.g., Coles, 2001; Méndez et al., 2006).

Moreover, the inter-annual variation and variations that are explained with covariables (e.g., climatic indices) can be incorporated in the distribution function in a manner similar to the way in which seasonal variation is incorporated (e.g. Izaguirre et al., 2010). For each parameter θ a series of covariables $C_i(t)$ and interannual variation of period T_j are introduced,

$$\theta(t) = \theta_0 + \sum_{k=1}^{N_k}(\theta_{ak}\cos 2k\pi t + \theta_{bk}\sin 2k\pi t) + \sum_{j=1}^{N_j}\left(\theta_{aj}\cos\frac{2\pi t}{T_j} + \theta_{bj}\sin\frac{2\pi t}{T_j}\right) + \sum_{i=1}^{N_i} f_i(C_i(t),t) \quad (10)$$

where long-term trends and other non-cyclic components are included as particular cases of the functions $f_i(C_i(t),t)$ in which there is no dependence on any covariable.

To evaluate the significance of the improvement in fit obtained when the order of the Fourier series is increased in any of the parameters, the Bayesian Information Criterion (BIC) can be used (Schwarz, 1978). BIC is given by (11)

$$BIC = -2\log(LLF) + \log(N_d)\,p \quad (11)$$

where LLF is the likelihood function, N_d is the number of available observations, and p is the number of model parameters. As long as the BIC is reduced, the increase on the number of parameters is considered to produce a significant improvement of the fit.

For the particular case of the mixture distribution model (1), Solari and Losada (2011) replaced the non-stationary parameters $u_1(t)$ and $u_2(t)$ by z_1 and z_2, using $F_c(u_1|t) = \Phi(z_1)$ and $F_c(u_2|t) = \Phi(z_2)$, where Φ is the standard normal distribution and z_1 and z_2 are stationary. In this way only the two stationary parameters z_1 and z_2 are estimated instead of the non-stationary thresholds $u_1(t)$ and $u_2(t)$; however, because the parameters of the central distribution $F_c(x|t)$ are non-stationary, the thresholds $u_1(t)$ and $u_2(t)$ are non-stationary as well. This way they achieved a significant reduction on the number of parameter that needs to be estimated when fitting the non-stationary version of model (1).

3. Time Series Simulation of Met-ocean Variables

At the present time there are, at least, two right paths to simulate time series of met-ocean variables or other geophysical variables: one that focuses on simulating storms and another that simulates complete series of values.

The method most widely used to simulate storms involves developing joint or conditioned distributions for the random variables of storm occurrence, intensity, and duration (Baquerizo and Losada, 2008). Based on these distributions, new time series are simulated assuming a standard shape for the storm. In general, storm occurrence is modeled using a Poisson distribution and storm intensity using a GPD; in addition it is common to condition the duration of a storm to its intensity (De Michele et al., 2007; Payo et al., 2008; Callaghan et al., 2008).This approach has been used for both uni- and multi-variate time series simulation. Although stationary functions are generally used for this purpose, non-stationary functions can also be employed, such as those proposed by Luceño et al. (2006), Méndez et al. (2008, 2006), and Izaguirre et al. (2010). A less frequent alternative in storm simulation is to assume that it is a Markov process and to use a multivariate distribution of extremes to model the time dependence of the variable while the storm lasts (Coles, 2001, Chap 8). This is the technique used by Smith et al. (1997), Fawcett and Walshaw (2006), and Ribatet et al. (2009).

Monbet et al. (2007) review methods for the simulation of complete series of values applied to wind and waves. The methods currently used can be classified as parametric and non-parametric.

The Translated Gaussian Process method (Borgman and Scheffner, 1991; Scheffner and Borgman, 1992; Walton and Borgman, 1990) is the most widely used nonparametric method. This method uses the spectrum of the normalized variable. Other nonparametric methods are those based on resampling, although according to Monbet et al. (2007) these methods are less frequently used.

The most frequently used parametric methods are based on autoregressive models. Studies employing such methods include Guedes Soares and Ferreira (1996), Guedes Soares et al. (1996), Scotto and Guedes Soares (2000), Stefanakos (1999) and Stefanakos and Athanassoulis (2001) for univariate series; for multivariate series, relevant studies include Guedes Soares and Cunha (2000), Stefanakos and Athanassoulis (2003), Stefanakos and Belibassakis (2005) and Cai et al. (2008). As in the Translated Gaussian Process method, before autoregressive models can be used the series must be normalized. For this purpose, non-stationary models of the mean and the standard deviation, like those proposed by Athanassoulis and Stefanakos (1995) Stefanakos (1999) and Stefanakos et al. (2006), are used. Recently Guanche et al. (2013) proposed a methodology for simulating multivariate time series based on the use of autoregressive and logistic autoregressive models, using daily mean sea level pressure fields as covariables. In this case a set of non-parametric PDF are used for the normalization of each variable.

The cited methods present the following limitations:

(a) Methods of normalizing variables are either stationary (e.g. Cai et al., 2008) or non-stationary. However, the latter ones focus on the center of the data distribution, generally using the non-stationary mean and standard deviation for normalization (e.g. Athanassoulis and Stefanakos, 1995; Guedes Soares et al., 1996).

(b) Parametric time dependence models are linear (e.g. Guedes Soares et al., 1996), piecewise linear (e.g. Scotto and Guedes Soares, 2000), or non-linear but are limited to the extremes (e.g. Smith et al., 1997).

(c) Generally speaking, the simulation is only evaluated using the mean, the standard deviation and the autocorrelation.

The method proposed by Solari and Losada (2011, 2014) and Solari and van Gelder (2012) to simulate nonstationary uni- and multi-variate time series with time dependence involves the use of a non-stationary parametric mixture distribution for the univariate distribution of the variables and of copulas or autoregressive models to model the auto- and cross-correlation of the variables. The proposed methodology is given next (§3.1). Afterward, univariate and multivariate dependence models for modelling auto- and cross-correlation of the variables are described (§3.2 and §3.3).

3.1. *Methodology*

The proposed methodology comprises three steps:

1) *Fitting non-stationary distributions functions and normalization of the variables*. For each one of the variables under study $\{V_i\}$ a non-stationary

distribution function $F_i(v_i(t)|t)$, as shown in §2, is fitted. Using these functions, variables are either transformed to a series of probabilities (12), with uniform stationary distribution, or normalized (13)

$$P_i = F_i(v_i(t)|t) \text{ being } P_i \sim \mathcal{U}(0,1) \tag{12}$$

$$Z_i = \Phi(P_i) \quad \text{ being } Z_i \sim \mathcal{N}(0,1) \tag{13}$$

where $\Phi(x)$ is the standard normal cumulative distribution function.

2) *Auto- and cross-crorrelation models fitting.* Short-term dependence models are fitted to represent the auto- and cross-correlation of the new stationary variables P_i or Z_i. To this end autoregressive models (§3.2.1 and §3.3.1) and copula-based Markov models (§3.2.2) are used.

3) *Time series simulation.* The simulation of new time series comprises two steps: first new time series of the stationary variables (P_i or Z_i) are simulated, and secondly these time series are transformed to the original variables by means of the non-stationery distributions $F_i(v_i(t)|t)$.

3.2. *Univariate Short-term Dependence Models*

3.2.1. *Autoregressive models*

Guedes Soares and Ferreira (1996) and Guedes Soares et al. (1996) applied univariate autoregressive models for the simulation of significant wave heights. Scotto and Guedes Soares (2000) extended the method by using univariate self existing threshold autoregressive models. More recently Cai et al. (2008) analyzed the use of autorregressive models for the simulation of time series of environmental variables.

An autoregressive-moving average $ARMA(p,q)$ model is given by (14)

$$Z(t) = \phi_1 Z(t-1) + \cdots + \phi_p Z(t-p) + \epsilon(t) + \theta_1 \epsilon(t-1) + \cdots + \theta_q \epsilon(t-q) \tag{14}$$

where, $\{\phi_i; i = 1, \dots, p\}$ and $\{\theta_j; j = 1, \dots, q\}$ are the coefficients of the autoregressive and of the moving average components of the model, respectively, and $\epsilon(t)$ stands for the independent, identically distributed realizations with a null mean and variance σ_ϵ^2 (a normal distribution is generally assumed for this). The $AR(p)$ model corresponds to the $ARMA(p,0)$ case.

To estimate the parameters of the $ARMA$ model, firstly the original variable $\{V(t)\}$ is transformed into a series $\{Z(t)\}$ via its probability distribution $F(V(t)|t)$, obtained using one of the models described in §2, and applying a standard normal distribution, as presented on (12) and (13). Once $\{Z(t)\}$ is obtained, the parameters $\{\phi_i; i = 1, \ldots, p\}$, $\{\theta_j; j = 1, \ldots, q\}$ and σ_ϵ^2 can be estimated using maximum likelihood estimation, least square or any other estimation procedure.

Once the model (14) is fitted, white noise is generated with variance σ_ϵ^2, and a new series $\{Z(t)\}$ is simulated using parameters ϕ_i and θ_j. Next, the series $\{Z(t)\}$ is transformed into $\{P(t)\}$, using the inverse of the standard normal distribution $P(t) = \Phi^{-1}(Z(t))$, and afterwards into the original variable $\{V(t)\}$, applying the inverse of its probability distribution $V(t) = F^{-1}(P(t)|t)$.

For the case of circular variables whose behavior cannot be approximated as linear, some special consideration may be required to fit autoregressive models. The interested reader is referred to Fisher (1993) and reference therein for further information on this respect.

3.2.2. *Copula-based Markov models*

A copula is a function $C: [0,1] \times [0,1] \to [0,1]$ such that for all $u, v \in [0,1]$, it holds that $C(0,1) = 0$, $C(u, 1) = u$, $C(0, v) = 0$ and $C(1, v) = v$; and for all $u_1 \leq u_2$, $v_1 \leq v_2 \in [0,1]$ it holds that

$$C(u_2, v_2) - C(u_2, v_1) - C(u_1, v_2) + C(u_1, v_1) \geq 0 \qquad (15)$$

The use of copulas to define multivariate distribution functions is based on the Sklar's theorem: when $F_{XY}(x, y)$ is a two-dimensional distribution function with marginal distribution functions $F_X(x)$ and $F_Y(y)$, there is then a copula $C(u, v)$ such that $F_{XY}(x, y) = Prob[X \leq x, Y \leq y] = C(F_X(x), F_Y(y))$.

For an introduction to copula theory the reader is referred to Joe (1997), Nelsen (2006) and Salvadori et al. (2007). Application of copula theory to marine climate and met-ocean variables can be found in e.g. De Michele et al. (2007), de Waal et al. (2007), Nai et al. (2004), Serinaldi and Grimaldi (2007) and Stefanakos (1999), among others.

Solari and Losada (2011) transform the original variable $\{V(t)\}$ into a uniformly distributed stationary variable $\{P(t)\}$ by means of (11). Next, following Abegaz and Naik-Nimbalkar (2008a,b) copula's theory is used to model the joint distribution of k successive states $(P(t), P(t-1), \ldots, P(t-k+1))$, i.e. to model a Markov process of order $k-1$. The joint probability $Prob[P(t), P(t-1)]$ of two consecutive values is represented by copula C_{12} such that

$$C_{12}(u, v) = Prob[P(t) \leq u, P(t-1) \leq v] \tag{16}$$

On this basis, the conditioned probability function is obtained. This function defines the distribution of $P(t)$ given $P(t-1)$ (or vice versa) and thus defines the first-order Markov process

$$C_{1|2}(u, v) = Prob[P(t) \leq u | P(t-1) = v] = \frac{\partial C_{12}}{\partial v}(u, v) \tag{17}$$

To define a Markov model of a higher order than one a copula construction process is used (Joe, 1997, chap. 4.5). This procedure is described below.

Given copula $C_{1\ldots k}$ (which defines the joint probability of k successive states) and, consequently, given the Markov model of order $k-1$, variables $F_{1|2\ldots k} = Prob[P(t)|P(t-1), \ldots, P(t-k+1)]$, and $F_{k+1|2\ldots k} = Prob[P(t-k)|P(t-1), \ldots, P(t-k+1)]$ are constructed. The dependence between the two variables is measured using Kendall's τ_k or Spearman's ρ_s. If this dependence is significant, then there is a relationship of dependence between $P(t)$ and $P(t-k)$ that cannot be explained by the Markov model of order $k-1$. In this case, it is necessary to construct a k-order Markov model. This can be accomplished using copula $C_{1\ldots k+1}$

$$C_{1\ldots k+1}(u_1, \ldots, u_{k+1}) = Prob[P(t) \leq u_1, \ldots, P(t-k) \leq u_{k+1}]$$
$$= \int_{-\infty}^{u_2} \cdots \int_{-\infty}^{u_k} C_{1k+1}\left(F_{1|2\ldots k}, F_{k+1|2\ldots k}\right) C_{2\ldots k}(dx_2, \ldots, dx_k) \tag{18}$$

Where C_{1k+1} is a bivariate copula fitted to the variables $F_{1|2\ldots k}$ and $F_{k+1|2\ldots k}$. This procedure is repeated until the value of k at which the dependence between variables $F_{1|2\ldots k}$ and $F_{k+1|2\ldots k}$ is not significant.

The procedure described is used to define multivariate copulas (i.e., those higher than the second order) based on a set of bivariate (i.e., second-order) copulas. Appendix B describes how this procedure is used to construct copula C_{1234}, which defines a third-order Markov process. The simulation process consists of two parts. First, the time-dependence model of copulas (18) is used to obtain a time series of probabilities $\{P(t)\}$; then, the non-stationary distribution of the original variable is used to obtain the variable time series from the probabilities series: $V(t) = F^{-1}(P(t)|t)$. To simulate the realization $P(t)$ of the Markov process of order $k-1$, once the previous realizations $P(t-1)$ to $P(t-k+1)$ are known, $u(t) \sim \mathcal{U}(0,1)$ is simulated and $P(t)$ obtained resolving equation (19)

$$u(t) = \frac{\partial C_{1\ldots k}}{\partial u_1 \ldots \partial u_k}(P(t), \ldots, P(t-k+1))$$
$$= \frac{\partial C_{1k}}{\partial F_{k|2\ldots k-1}}\Big(F_{1|2\ldots k-1}(P(t), \ldots, P(t-k+2)), \tag{19}$$
$$F_{k|2\ldots k-1}(P(t-1), \ldots, P(t-k+1))\Big)$$

where C_{1k} is the bivariate copula fitted to $F_{1|2\ldots k-1}$ and $F_{k|2\ldots k-1}$ to construct $C_{1\ldots k}$ and where to $F_{1|2\ldots k-1}$ and $F_{k|2\ldots k-1}$are calculated using the set of bivariate copulas $C_{1k-1}, C_{1k-2}, \ldots, C_{12}$.

When this procedure is used, it is not necessary to use equation (18) to perform the simulations because equation (19) can be solved using only the bivariate copulas. To obtain $P(t)$, equation (19) can be numerically solved using the bisection method. The simulation process for a third-order Markov model is described in Appendix C.

In §4 two copula families are used in a study case. A list of copula families, their characteristics, and the different ways to fit them to the data can be found in the works of Joe (1997), Nelsen (2006) and Salvadori et al. (2007). For a summary of methods and goodness-of-fit tests, see Genest and Favre (2007) and references therein.

3.3. *Multivariate Short-Term Dependence Models*

3.3.1. *Vector-Autoregressive*

Guedes Soares and Cunha (2000) used bivariate autoregressive models for the simulation of multivariate time series, namely wave height and peak period. Stefanakos and Belibassakis (2005) use a vector autoregressive moving average model for the simulation of wave height, peak period and wind speed. Cai et al. (2008) applied a bivariate AR model for the study of wave heights and storm surges, and more recently Guanche et al. (2013) used a combination of autoregressive and logistic autoregressive models to simulate trivariate sea states (wave height, period and direction).

Autoregressive models give the value of the current observation as a linear function of past observations and a white noise. Vector Autoregressive models are an extension of autoregressive models for multivariate data. For a description of vector autoregressive models the reader is referred to Lütkepohl (2005). The Vector Autoregressive model $VAR(p)$ of order p is given by

$$Y(t) = V + \sum_{i=1}^{p} A_i Y(t-i) + U(t) \tag{20}$$

where $Y(t) = \left(Y_1(t), \dots, Y_K(t)\right)^T$ is a vector of dimensions $(K \times 1)$, being K the number of variables; each A_i is a matrix of autoregressive coefficients of dimensions $(K \times K)$, $V = (V_1, \dots, V_K)^T$ is a vector of dimension $(K \times 1)$ that allows for a non-zero mean for the variables $\mathbb{E}\left(Y(t)\right)$, being $\mathbb{E}(\cdot)$ the expected value; and $U(t) = \left(U_1(t), \dots, U_K(t)\right)^T$ is a K -dimensional white noise process, also called innovation process or error, that must fulfill $\mathbb{E}\left(U(t)\right) = 0$, $\mathbb{E}(U(t)U(t)^T) = \Sigma_U$ and $\mathbb{E}(U(t)U(s)^T) = 0$ for $t \neq s$.

In this work the parameters of the $VAR(p)$ model are estimated through Least Square (see Lütkepohl 2005, Ch. 3). For this the model is expressed on matrix notation as $\mathbb{Y} = \mathbb{B}\mathbb{Z} + \mathbb{U}$, where $\mathbb{Y} = (Y_1, \dots, Y_T)$, $\mathbb{B} = (A_1, \dots, A_p)$, $\mathbb{Z} = (Z_0, \dots, Z_{T-1})$ with $Z_t = \left(1, Y(t), \dots, Y(t-p+1)\right)^T$ and $\mathbb{U} = (U_1, \dots, U_T)$, being T the number of observations available for estimation. Then, autoregressive parameters are estimated as $\widehat{\mathbb{B}} = \mathbb{Y}\mathbb{Z}^{-1}(\mathbb{Z}\mathbb{Z}^T)^{-1}$; while the covariance matrix Σ_U of the withe noise $U(t)$ is estimated through the errors $\widehat{\mathbb{U}} = \mathbb{Y} - \widehat{\mathbb{B}}\mathbb{Z}$ as $\widehat{\Sigma_U} = \widehat{\mathbb{U}}\widehat{\mathbb{U}}^T/(T - Kp + 1)$.

For defining the order p of the model that should be used it is possible to use the Bayesian Information Criteria (11). Procedure is as follows: first model parameters are estimated for a series of orders $\{p\}$; then, LLF and BIC are estimated for each one of the models and the one with the lower BIC is selected as the "optimum" model.

Assuming that the white noise follows a multivariate normal distribution of zero mean and covariance $\widehat{\Sigma_U}$, the LLF is

$$LLF = \sum_{t=1}^{T} \log\left(f_{MVN}\left(\widehat{U}(t) \middle| 0, \widehat{\Sigma_U}\right)\right) \tag{21}$$

where $f_{MVN}\left(\widehat{U}(t) \middle| 0, \widehat{\Sigma_U}\right)$ is the density function of the multivariate normal distribution.

Other, more sophisticated VAR models, are the Threshold VAR ($TVAR$) and the Markov Swithcing VAR ($MSVAR$) models. $TVAR$ model assume that there exists more than one possible regime for the system, i.e. more than one VAR model, and that at each time t the regime is defined by the value taken by the variable z at time $t - d$, where d is the delay. When z is one of the variables of the regression, the model is called Self Exiting. In the $MSVAR$ model it is assumed the existence of an unobserved variable $s(t)$ that determines the regime at each time steps, and that this variables follows a discrete Markov process. Again, for each regime a VAR model is defined. For examples of applications of $TVAR$ and $MSVAR$ models to the simulation of met-ocean variables the reader is referred to Solari and van Gelder (2012) and references therein.

4. Applications

4.1. *Univariate Stationary Analysis: Significant Wave Heights at Gulf of Cádiz*

Here a series of hindcast spectral significant wave height, provided by Puertos del Estado, Spain (www.puertos.es), corresponding to the Gulf of Cádiz (36.5°N, 7.0°W) is used for exemplifying the use of the stationary mixture model (1) presented on §2.1.1. The location of the data point is shown in Fig. 1.

The Cádiz series comprises 13 years and 3 months of sea states with a duration of 3 h, although with some gaps in the record (36,496 data). The most severe storms in The Gulf of Cádiz occur during the passage of low pressure systems from the Atlantic Ocean to the Mediterranean Sea. Wind uses to blow form the W–SW with an average speed of 25 m/s over a fetch of more than 400 km.

The fitting of the mixture model is presented on §4.1.1. Then the use of the mixture model as a tool for threshold estimation for applying the POT method is presented on §4.1.2. Finally, on §4.1.3, the results are discussed.

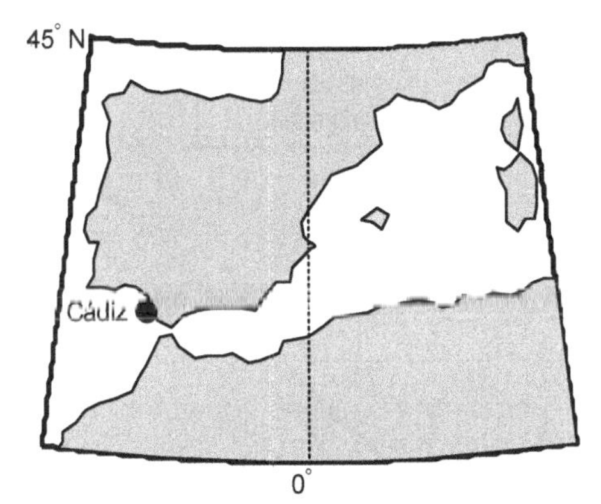

Figure 1. Location of the Gulf of Cádiz data point.

4.1.1. *Mixture models*

The usual distributions for modeling the central regime of significant wave height are the Log-Normal (LN), Weibull, Gamma, etc., distributions, although the one that provides the best fit for the data series in this case is the LN, justifying the adoption of this distribution for the central regime in model (1), called LN-GPD distribution below. Furthermore, in this study case, the LN distribution was used as a reference to evaluate the performance of the LN-GPD model.

The estimated values of the LN-GPD parameters are listed in Table 1 along with their standard deviations (details on estimation procedure are found in Appendix A and Solari and Losada 2012a). Fig. 2 shows the empirical cumulative distribution function (CDF), along with the functions obtained with the fitted LN-GPD mixture model (1) and with the LN distribution. It can be observed that the LN-GPD distribution improves the fit in the tails with respect to the fit obtained with the LN distribution.

Table 1. LN-GPD parameters and their corresponding standrd deviation

Parameter	u_1	u_2	μ_{LN}	σ_{LN}	ξ_2
Estimated value	0.41	3.7	-0.103	0.616	-0.113
and standard deviation	3.6×10^{-3}	0.081	3.2×10^{-3}	2.3×10^{-3}	0.028

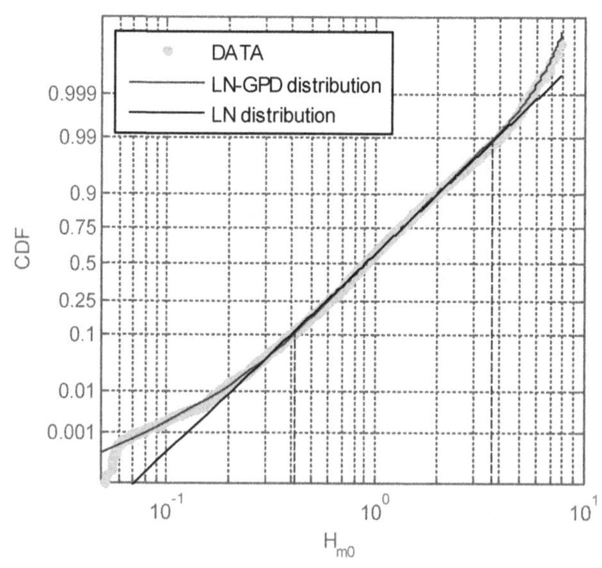

Figure 2. Empirical (gray dots), LN (black line) and LN-GPD (blue line) CDF for the Cádiz data series. Thresholds u1 and u2 in dashed lines.

4.1.2. *Extreme value analysis*

The use of the upper threshold u_2, obtained from the LN-GPD distribution, in the application of the POT method is compare with other proposed methods as follows:

(a) The upper threshold necessary to apply the POT method is estimated by using other available methodologies, namely, the following: the graphical methods presented in Coles (2001) and the methods that have recently been presented by Thompson et al. (2009) and by Mazas and Hamm (2011).
(b) The extreme value distribution is obtained for each of the thresholds, comparing the quantiles of the high-return period and their confidence intervals.

In all of the cases, to define the storms, a minimum time of 2 days is imposed between the end of an excess over the threshold and the beginning of another. Sometimes this results in that fewer peaks per year are obtained using lower thresholds than using higher thresholds. Another way of defining storms could result in different behavior, but the study of the sensitivity of the POT method to this parameter is beyond the scope of this work.

The upper threshold obtained from the LN-GPD distribution is $u = 3.7\ m$. Fig. 3 presents the two plots required for estimating the threshold with the graphical method: the MRLP and graphs of ξ and σ^* (see §2.2.1). The threshold value selected based on Fig. 3 is $u = 3.5\ m$ (indicated in Fig. 3 by a red circle).

Table 2 summarizes the result of applying the method proposed by Thompson et al. (2009) using different values of significance α for the hypothesis test and a series of thresholds constructed with different numbers of elements N and upper thresholds u_{max}. To select the threshold, the criterion was selected of using the value obtained using a significance of $\alpha = 0.2$ for the chi-square test and a series of $N = 100$ thresholds with the upper threshold u_{max} corresponding to the 98% percentile, as recommended by Thompson et al. (2009). With this criterion, $u = 2.5\ m$ is obtained.

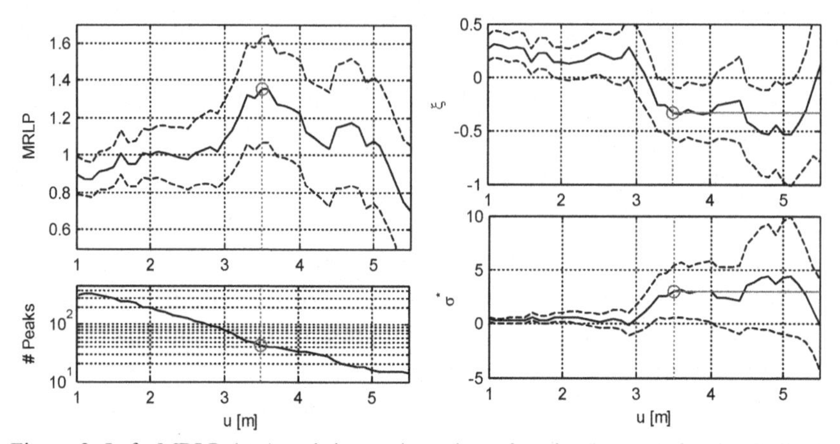

Figure 3. Left: MRLP (top) and the total number of peaks (bottom) for the Cádiz data series. Right: evolution of $\boldsymbol{\xi}$ (top) and $\boldsymbol{\sigma}^*$ (bottom) for different thresholds for Cádiz data series. The chosen threshold is indicated with red in the four panels.

Table 2. Upper thresholds obtained by applying the methodology proposed by Thompson et al. (2009) using different numbers of elements ($\boldsymbol{N}$) and upper thresholds ($\boldsymbol{u_{max}}$) for the definition of the threshold series and different significance for the chi-square test ($\boldsymbol{\alpha}$).

	$\alpha = 0.2$		$\alpha = 0.05$	
u_{max}	*N=100*	*N=500*	*N=100*	*N=500*
98.0% (3.3 m)	2.5	3.1	1.3	3.1
98.5% (3.6 m)	2.6	3.4	2.2	3.4
99.0% (3.9 m)	2.3	3.8	1.7	3.8
99.4% (4.4 m)	0.9	4.1	0.9	4.1

The threshold obtained with the LN-GPD model ($u = 3.7\ m$) results in 3.4 peaks per year, whereas the threshold selected using the graphical method ($u = 3.5\ m$) results in 3.6 peaks per year, and the one identified by the method of Thompson et al. (2009) ($u = 2.5\ m$) results in 10.8 peaks per year. The thresholds selected with the graphical method and with LN-GPD result in a number of peaks per year that is within the range recommended by Mazas and Hamm (2011), whereas the threshold obtained using the method of Thompson et al. (2009) exceeds this range.

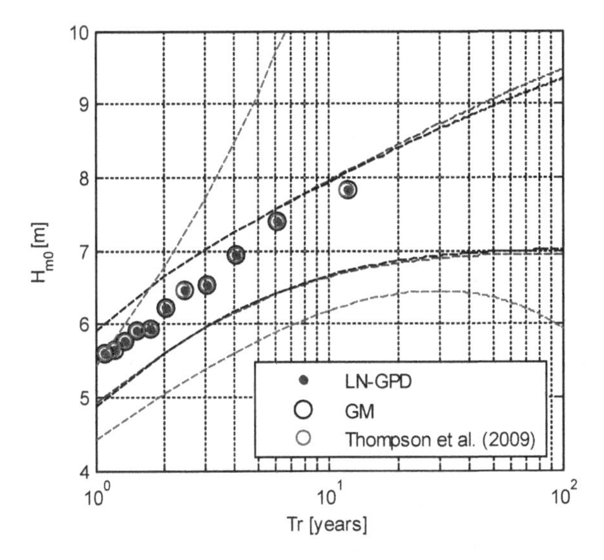

Figure 4. Peak over threshold series and 90% confidence intervals for the quantiles obtained with the GPDs fit using thresholds 3.7 m (blue), 3.5 m (black) and 2.5 m (red). GM stands for graphical method.

Next, a GPD distribution was fit to the POT data series constructed using the different thresholds defined previously. The 90% confidence intervals of the quantiles estimated for various return periods are shown in Fig. 4. The figure includes the empirical quantiles obtained with the different POT data series. Table 3 presents the values of significant wave height from a 100-year return period obtained with the different thresholds, along with the corresponding confidence intervals.

Table 3. 100-years return period significant wave height ($\boldsymbol{H_{s,Tr=100}}$) obtained with different thresholds and their correponding 90% confidence intervals (CI).

Method	Threshold [m]	$H_{s,Tr=100}$ [m] (90% CI)
LN-GPD	3.7	8.2 (7.0-9.5)
Graphical method	3.5	8.2 (7.0-9.3)
Thompson et al. (2009)	2.5	15.3 (5.9-24.7)

4.1.3. *Discussion*

The mixture distribution LN-GPD achieved a significantly better fit of the data than the LN distribution. However, since the models increase the number of parameters from 2 with the LN to 5 with LN-GPD, it was necessary to verify that the improvement of the fit was significant enough to justify this increase in the number of parameters. For that purpose, BIC (11) was estimated for both distributions, resulting in a lower BIC for the LN-GPD than for the LN, i.e. the increase on the number of parameter is justified by the improvement on the fit.

Additionally, the threshold estimated by the LN-GPD proves to be appropriate to apply the POT method and to calculate high-return period quantiles. Also, the LN-GPD model provided the upper threshold as a parameter of the model, which meant that it could be used as an objective and automatic method for threshold estimation. However, one disadvantage of using the LN-GPD is that it requires the likelihood function to be written based on the central distribution f_c to be used for modeling the data, whereas the other methods can be programmed independently of the data.

For a detail discussion of the results obtained with the four methods used to select the threshold, namely the graphical method, the methods proposed by Thompson et al. (2009) and by Mazas and Hamm (2011), and the use of parameter u_2 obtained by fitting LN-GPD, the reader is referred to Solari and Losada (2012a).

4.2. *Univariate Non-Stationary Analysis and Time Series Simulation*

In §4.2.1 the significant wave height data used on §4.1 is fitted with a non-stationary version of the LN-GPD model, called NS-LN-GPD from now on, as described on §2.3. Then the model is used for applying the time series simulation methodology described on §3.1. To this end both of the dependence models described on §3.2 are fitted on §4.2.2, namely the autoregressive models and the copula-based Markov model. Afterward, on §4.2.3 new series are simulated and compared with the original ones. Finally, results are discussed on §4.2.4.

4.2.1. *Non-stationary distribution*

In this section the NS-LN-GPD parameters are estimated. A non-stationary LN distribution (NS-LN) is also fitted (corresponding to the NS-LN-GPD with Z_1 and Z_2 parameters approaching infinity) for use in testing the goodness of fit obtained using the NS-LN-GPD model.

In the first instance, the parameters are only allowed to have seasonal variations (i.e., variation of periods less than or equal to a year (9)); interannual variation, covariables and trends were not considered. Different models are fitted to the data with the Fourier series having a maximum order of approximation N_k varying between 1 and 4. The order 1 represents annual variation, 2 represents semiannual variation, and so on. For each fitted model, the BIC (11) is estimated.

Each model is identified using three digits $[abc]$; a is the order of approximation of the Fourier series used for μ_{LN}, b is the order of approximation of the series used for σ_{LN}, and c is the order of approximation of the series used for ξ_2. When a maximum approximation N_k is allowed, $a, b, c \leq N_k$ should hold. The total number of parameters of the model $[abc]$ is $2(a + b + c) + 5$; i.e., there are $2a + 1$ parameters to be used in the Fourier series representation of μ_{LN}, $2b + 1$ parameters to be used in the Fourier series representation of σ_{LN}, $2c + 1$ parameters to be used in the Fourier series representation of ξ_2, and the two stationary parameters Z_1 and Z_2.

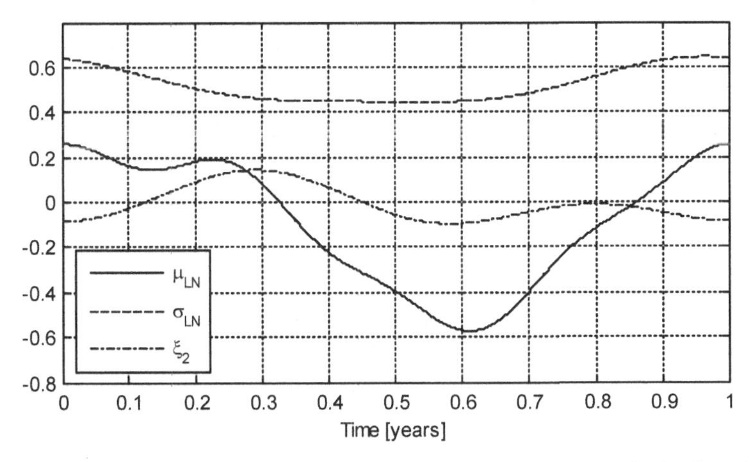

Figure 5. Time evolution of $\boldsymbol{\mu_{LN}}$, $\boldsymbol{\sigma_{LN}}$ and $\boldsymbol{\xi_2}$ for the NS-LN-GPD [4, 2, 2] model

The model with the minimum BIC in this case is [4 2 2]: i.e., a Fourier series of order 4 for μ_{LN} and of order 2 for σ_{LN} and ξ_2. Fig. 5 shows the annual temporal evolution of these parameters. As can be observed, the principal component is the annual period, and the other components provide non-negligible corrections of a lesser order. The only exception is parameter ξ_2, for which the semiannual component is of the same order of magnitude as the annual one.

The fit of the [4 2 2] model obtained using the NS-LN-GPD parameters is compared with that of the model obtained using the NS-LN (also using order 4 for the Fourier series). Fig. 6 shows the quantiles corresponding to the empirical accumulated probability values and those obtained when the NS-LN and NS-LN-GPD models are used. The empirical quantiles have been obtained using a moving window of one month. Generally speaking, the quantiles calculated using the NS-LN-GPD distribution coincide with the empirical quantiles. As compared with the NS-LN model, the NS-LN-GPD model exhibits superior fit at the tails. Fig. 7 shows the annual CDF on log-normal paper and the annual PDF for both non-stationary models and for the original data. As can be observed, the NS-LN-GPD model exhibits a better fit than the NS-LN model, particularly at the tails, but also at the mode of the distribution.

When the moving average of the data is displayed on a graph (Fig. 8), not only the seasonal cycle but also two trends are observed: (i) a cyclical component with a period of approximately 5 years and (ii) a decreasing trend. To analyze both, the following cyclical components are included in μ_{LN} in order to show how the proposed model can include the interannual variations observed in the series:

$$\mu_{LN,annual} = a_{i1}\cos(2\pi t/5) + b_{i1}\sin(2\pi t/5) + a_{i2}\cos(2\pi t/26) + b_{i2}\sin(2\pi t/26) \quad (22)$$

This is an ad hoc model for long-term trends that assumes that the downward trend in the 13 years of data is part of a 26-year pattern of cyclical variation. It is not our aim to perform an in-depth analysis of the interannual variation in the data series being used; this would mean studying covariables of interest such as the North Atlantic Oscillation (NAO) and considering long-term trends and climate cycles, which require longer series than the one available as well as series of covariables (see e.g. Izaguirre et al., 2010; Ruggiero et al., 2010).

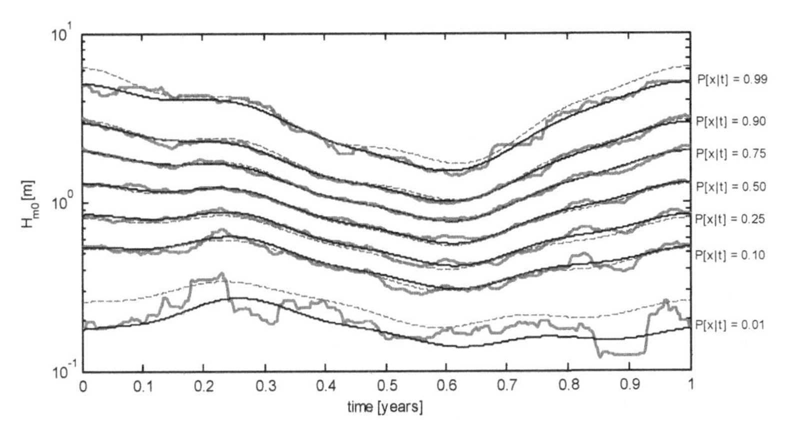

Figure 6. Iso-probability quantiles for non-exceeding probability P[x|t] equal to 0.01, 0.1, 0.25 0.5, 0.75, 0.9 and 0.99; empirical (grey continuous line), NS-LN model (red dashed line) and NS-LN-GPD model (black continuous line).

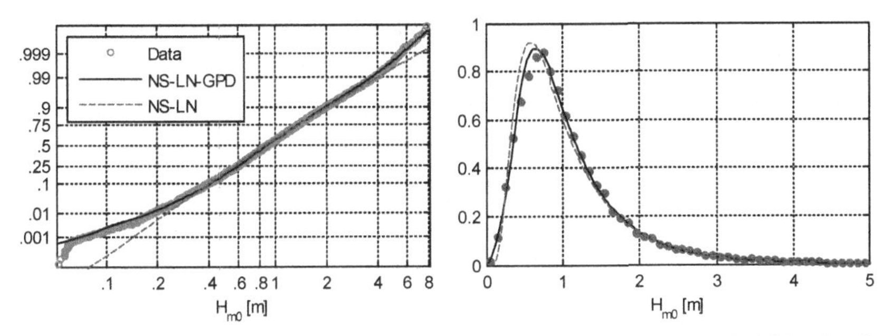

Figure 7. Accumulated probability on log-normal paper (left) and probability density (right). Empirical (dots), NS-LN normal model (dashed line), and NS-LN-GPD model (continuous line).

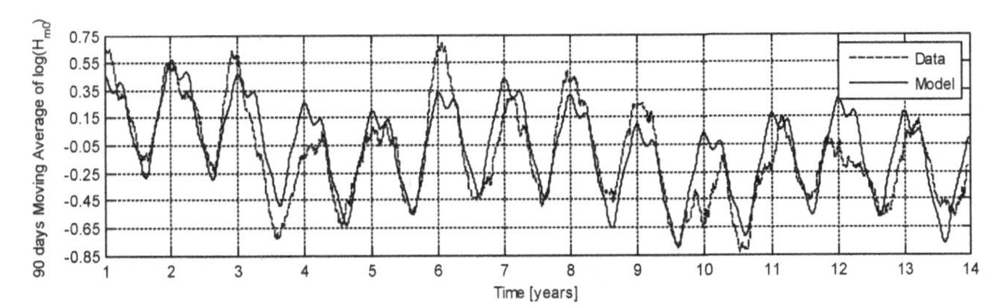

Figure 8. Ninety-day Moving Average of the significant wave height and the μ_{LN} parameter of the interannual model.

The four parameters introduced in (22) along with the other parameters of the model are estimated using maximum likelihood estimation with 4 as the maximum order of approximation for the Fourier series and using the BIC to select the model. The model obtained in this case is [4 2 2 2], where the first three numbers refer to the order of approximation of μ_{LN}, σ_{LN} and ξ_2 and the last refers to the two interannual cyclical components included in μ_{LN} (22). The evolution of μ_{LN} obtained with [4 2 2 2] model is presented on Fig. 8.

4.2.2. *Short-term dependence*

To fit the time dependency, different copula families can be tested. In this study, the families that best fit the data are selected based on the log-likelihood function (LLF) and a visual evaluation. The following paragraphs describe the data fitting processes, which are conducted based on the probability series $\{P(t)\}$ obtained by means of (12) using NS-LN-GPD model [4 2 2 2].

The asymmetric Gumbel-Hougaard copula (see e.g. Salvadori et al., 2007) provides a good fit for the time-dependence between $P(t)$ and $P(t-1)$. Fig. 9 depicts the empirical function $C(P(t), P(t-1))$ and that obtained using the asymmetric Gumbel-Hougaard function. It is clear that the modeled and empirical isoprobability curves overlap, except around $P(t) \approx P(t-1) \approx 0.1 - 0.4$, where the data reflect a more marked dependence than that exhibited by the model. In general, the fit is good.

Then the dependence between $P(t)$ and $P(t-2)$, which was not explained by $C_{12}(P(t), P(t-1))$, is estimated. For this purpose, the C_{12} copula was used to estimate $F_{1|2}$ and $F_{3|2}$. The dependence between $F_{1|2}$ and $F_{3|2}$ is significant ($\tau_K = -0.133$ and $\rho_S = -0.192$), and thus, the trivariate copula C_{123} was constructed.

To obtain the trivariate copula (B2), the bivariate copula $C_{13}(F_{1|3}, F_{3|2})$ was fitted. In this case, a good fit was obtained using the Fréchet family (see e.g. Salvadori et al., 2007), although there was some asymmetry in the data that was not captured by the copula (see Fig. 9).

The copula C_{123} was used to estimate $F_{1|23}$ and $F_{4|23}$. The dependence between these variables was found to be $\tau_K = -1.4 \times 10^{-3}$ and $\rho_S = -1.3 \times 10^{-4}$. Consequently, the variables $F_{1|23}$ and $F_{4|23}$ can be regarded as independent.

A high-order ARMA(p,q) model was also estimated to compare the results obtained with the copula based Markov model. An optimal number of parameters was not selected; rather a sufficiently high number ($p = q = 23$) was used to take advantage of the capacities of these models.

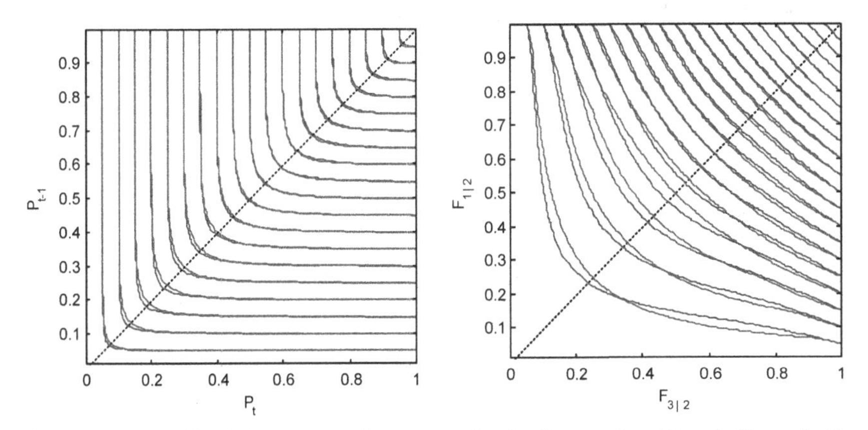

Figure 9. Empirical (green) and parametric (red) copulas C_{12} (left) and C_{13} (right).

4.2.3. *Simulation and verification*

A simulation was conducted of 500 years of significant wave height using the NS-LN-GPD (including the interannual variation terms (22)), along with: (a) the copula based Markov model (C-Model), and (b) the ARMA(23,23) model (A-Model).

Figure 10 shows a five-year data series and another five-year series simulated using the C-Model. Next, the results obtained using the different models are evaluated, differentiating between the medium or main-mass regime and the extreme or upper-tail regime.

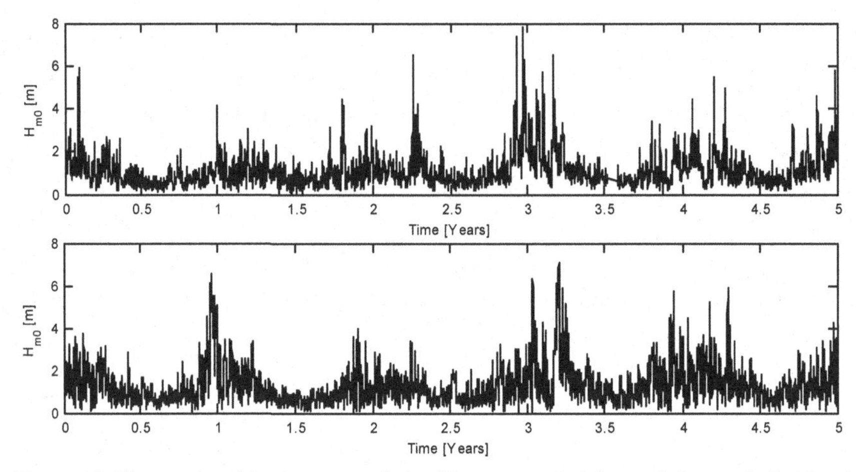

Figure 10. Five years of (top) measured significant wave heights and (bottom) simulated significant wave heights.

Medium or Main-Mass Regime

The medium regime obtained using both models (C- and A- Model) are very similar. In fact, it is practically impossible to differentiate between the two simulated series in the PDF and CDF plots. Since the simulated data series are very long, their distributions are equal to the theoretical distributions (NS-LN-GPD and NS-LN, respectively). As a consequence the comparison of the CDF and PDF obtained from the simulated and the original data series is similar to the results shown in Fig. 7.

Table 4 shows the values of the statistics derived from the first four moments of the distribution: mean, variance, skewness, and kurtosis. As can be observed, both models properly represent the mean and the variance. Regarding skewness and kurtosis, the best approximations were obtained using the C-Model, although both models yielded overestimated figures for kurtosis.

Table 4. Statistics obtained from the first four central moments

	Data	C-Model	A-Model
Mean	1.088	1.086	1.093
Variance	0.548	0.538	0.556
Skewness	2.127	2.275	2.410
Kurtosis	10.006	12.290	14.326

Figure 11 shows the autocorrelation function (ACF) for the data and the two simulated series. For a time lag of less than three days, the C-Model fits the data better than the A-Model. In contrast, for longer time-lags, the A-Model provides a better fit. The main reason for this is that the ARMA model is a 23rd -order model, whereas the copula-based models correspond to a second-order Markov model. When third-order ARMA models are used (as indicated by the red dashed line referred to as ARMA(3,3) in Fig. 11), the long-term fit of the ACF is equivalent to that obtained using copula-based models, whereas the short-term fit is roughly the same as that obtained using a 23rd -order ARMA model.

Figure 12 shows the PDF of the persistence over different thresholds (0.5 m, 1.0 m, 1.5 m, 2.0 m, 2.5 m, 3.0 m). In many cases, there are discrepancies between the persistence regimes for the original and simulated data series. For a threshold of 0.5 m, the simulated series show a lower than observed frequency of persistence of short duration (6 hours); i.e., the simulations overestimate persistence over 0.5 m. For thresholds greater than 2 m, the simulations (particularly those obtained using A-Model) show a higher than observed frequency of persistence of short duration (6 hours); i.e., both the C- and A-Models underestimate persistence, but the extent of the underestimation by the A-Model is greater. Nevertheless, for thresholds greater than 1.5 m, the series obtained using the C-Model shows a better fit with regard to the persistence than that obtained using the A-Model. In contrast, for the thresholds 0.5 m and 1 m, the data series simulated using the A-Model exhibits a better fit with regard to the persistence than the series simulated using the C-Model.

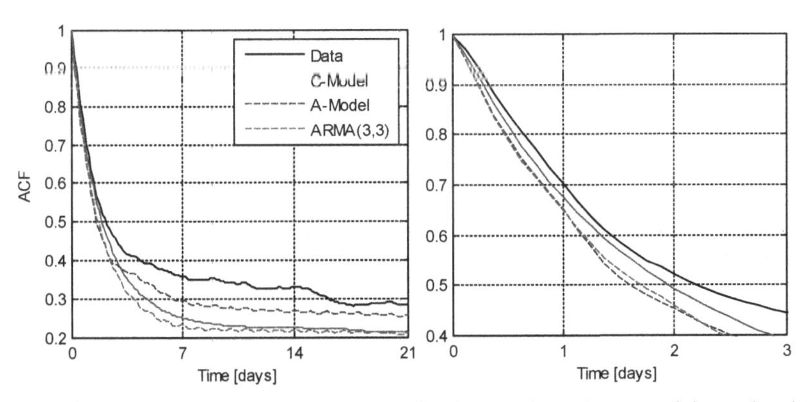

Figure 11. Autocorrelation function (ACF) for the two dependence models used and for a simulation run using an ARMA(3,3) model.

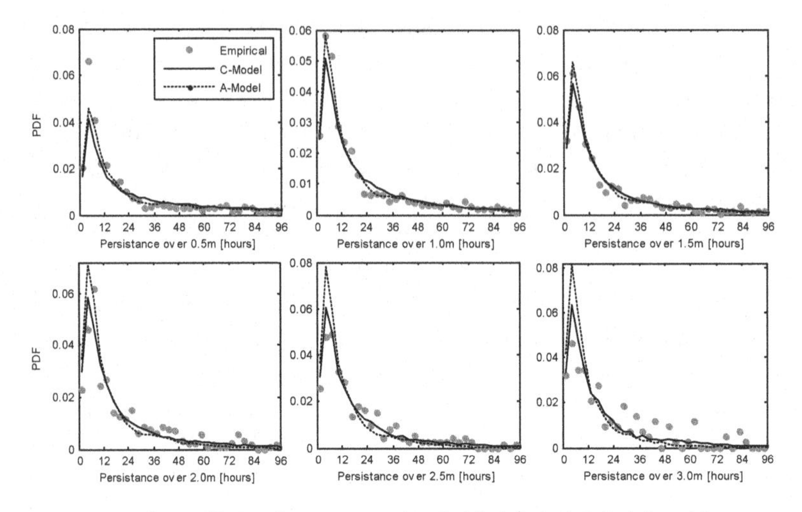

Figure 12. Persistence over thresholds 0.5, 1, 1.5, 2, 2.5 and 3 m.

Extreme or Upper-Tail Regime: Annual Maxima

Figure 13 shows the annual maxima of the empirical data and of the simulated series for different return periods. Generally speaking, the A-Model has overestimated the annual maxima, whereas the data obtained via the C-Model underestimates it. Nevertheless, the series simulated using the C-Model appropriately fit the empirical regime of annual maxima.

Extreme or Upper-Tail Regime: Storms and Peaks Over Threshold (POT)

Storms were identified following section §4.1.2: a threshold $u = 3.58\ m$ was used and the minimum time between the storms was 2 days. The mean number of storms per year based on these figures was $N_{Data} = 3.08$. The mean numbers of storms per year based on the simulated series were $N_{C-Model} = 3.46$ and $N_{A-Model} = 6.66$.

Figure 14 shows the values of significant wave height corresponding to different return periods as obtained from the POT regime. It also displays the fit of the GPD obtained in §4.1.2 for that regime. The series obtained using the A-Model significantly overestimates the data and reflects a long-term trend that is very different from the trend indicated by the GPD. On the other hand, although the C-Model underestimated the data for return periods of less than 10 years, the series obtained lies within the GPD confidence limits for high return periods.

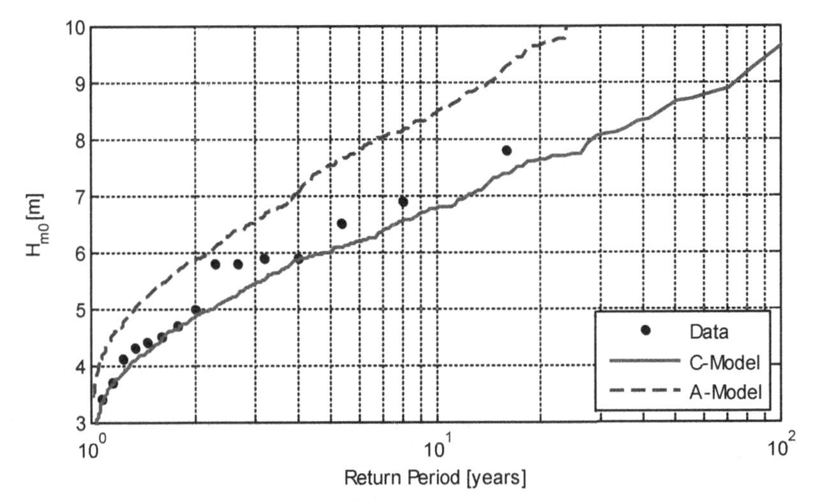

Figure 13. Annual maxima significant wave height: original data (dots), data from the C-Model (green continues line) and data from the A-Model (blue dashed line).

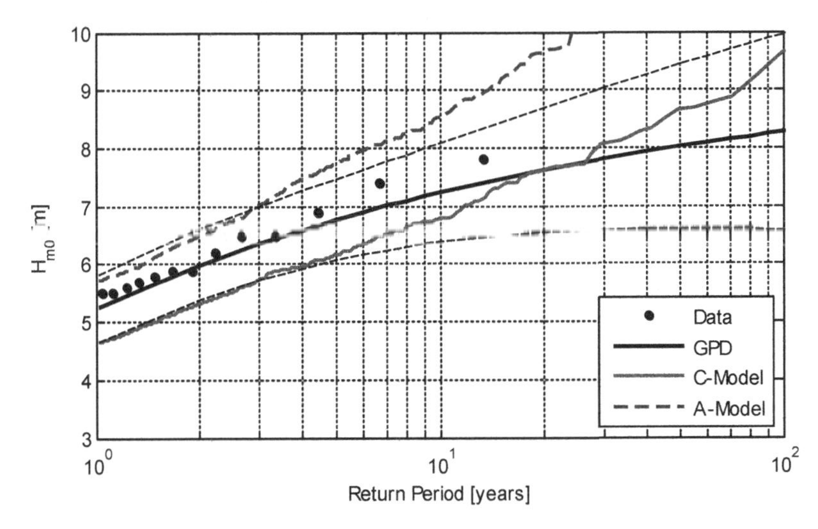

Figure 14. POT regime for significant wave height: original data (dots); GPD (continues black line) with confidence intervals (dashed black line) fitted to the original data; C-Model data (green continues line) and A-Model data (blue dashed line).

4.2.4. *Discussion*

With regard to the marginal distribution, all of the simulated series have approximated the original data quite well. The differences between the models become evident when the autocorrelation and persistence regimes are analyzed. As compared to the ARMA model, the copula-based time-dependency model provides a better fit to persistence data for thresholds higher than 1 m.

With respect to autocorrelation, it appears that in the long term (with time-lags longer than 3 days), the high-order autoregressive models (23) provide better fit to the data than do the models based on copulas. However, when low-order autoregressive models (of order 3) are used, the long-term behavior of the autocorrelation is similar to that obtained using copula-based models (which are also low-order models). If only short-term behavior is considered (with a time lag of less than 3 days), the copula-based models show a slightly better fit in terms of autocorrelation than that obtained using autoregressive models.

For the extreme regime the copulas-based model provided the best results in both the annual maxima and the POT analysis. The data from the ARMA-based models indicate that there was a much larger mean number of storms per year than was actually recorded. Although not presented here, the copula-based models also appropriately fit the recorded data regarding the seasonal distribution of the number of storms per year and their duration (see Solari and Losada, 2011).

Based on these findings, copula-based models can be deemed more suitable for use than are ARMA-based models given the frequency and persistence of the storms, which are important parameters to consider when studying systems such as beaches or ports.

4.3. ***Multivariate Non-Stationary Models and Time Series Simulation***

In this section the application of the multivariate simulation methods described on §3.3 is presented. In §4.3.1 a bivariate series of significant wave height and peak wave period is simulated using the VAR models introduced on §3.3.1. Then, results obtained are discussed on §4.3.2.

4.3.1. *VAR models*

A bivariate time series of significant wave height and peak period is simulated. The data used for fitting the models is the same used on the previous sections: hindcast wave data from the Gulf of Cadiz, Spain.

The parameters of the non-stationary marginal distributions of the two variables are obtained through maximum likelihood. For significant wave height the seasonal NS-LN-GPD model fitted in §4.2.1 is used. For the peak period a mixture model of two non-stationary log-normal distributions (NS-Bi-LN), as presented on §2.1.2, is used. The order of approximation of the Fourier series is varied between 1 and 4. For each model the BIC is estimated, and the one with the lower BIC is selected. The procedure is similar to that used in §4.2.1.

Models give a very good fitting for the two variables under study. NS-LN-GPD model fitted to the significant wave height data was evaluated on §4.2.1. Fig. 15 presents the results obtained by fitting the NS-Bi-LN model to the peak period series, showing the annual mean probability density function (PDF) and the non-stationary empirical and modeled quantiles. The agreement between the model and the data is noticeable.

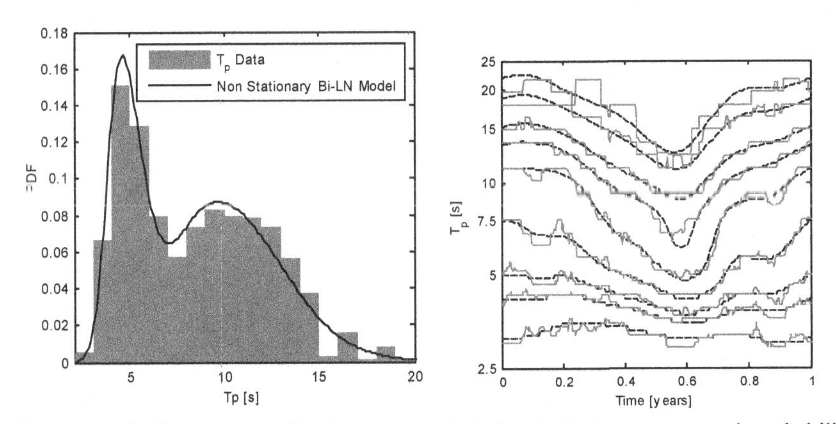

Figure 15. Left: empirical (bars) and modeled (black line) mean annual probability density function. Right: empirical (gray) and modeled (dashed black) quantiles of 1, 5, 10, 25, 50, 75, 90, 99 and 99.9% for Tp as a function of time.

These two distributions are used to normalize the variables using (12) and (13), obtaining a bivariate series of stationary standard normal variables $\{Z_{H_s}(t), Z_{T_p}(t)\}$. Then, a vector autoregressive (VAR) model is fitted to this bivariate series. The parameters of the $VAR(p)$ model are estimated through the least square method described on §3.3.1 for order p between 1 and 8. For each model the BIC is calculated, and the lower BIC is obtained with $p = 7$.

A series of 500 years of the variables $\left\{Z_{H_s}(t), Z_{T_p}(t)\right\}$ is simulated using the VAR(7) model. Then, these are transformed to the original variables $\{H_S(t), T_P(t)\}$ by means of the non-stationary marginal distributions NS-LN-GPD and NS-Bi-LN. Next, the simulated series is compared with the original one in terms of its marginal distributions (uni- and bi-variate) and its interannual variability.

In Fig. 16 the annual univariate PDF of each of the 500 years simulated with the VAR(7) model are presented, along with the PDF of the 13 measured years and its mean annual PDF. First, it is noticed that the simulated series are able to reproduce the mean annual PDF of the measured series. On the other hand, the simulations show a significant variation in the annual PDF. The annual PDF of the simulated series produce a cloud around the annual mean PDF of the measured data that includes most of the measured annual PDF. However, there are at least two years of measured data whose PDF cannot be reproduce by the simulated series. Those two years correspond to the first two years of the measured series, for which severer weather was observed, i.e. higher wave heights and wind speed and lower peak period.

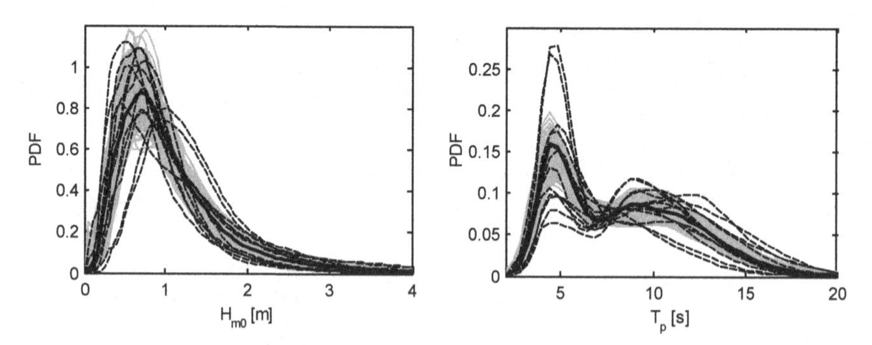

Figure 16. Interannual variability of the annual PDF of Hm0 (left) and Tp (right). Grey: annual PDF for each simulated year. Dashed black: annual PDF for each measured year. Continuous black: mean annual measured PDF.

It is important that the simulated series reproduce not only the marginal bivariate distributions of the measured data but also its marginal multivariate distributions, since the latter contain information about the joint occurrence of values of the variables. It was observed that data series simulated with VAR model reproduce well those bivariate distributions of the normalized variables whose behavior is similar to that of a multivariate normal distribution, i.e. with only one mode and with constant dependence structure for the whole range of values of the variables. When the bivariate distribution shows multimodality, as is the case of $\{Z_{H_s}(t), Z_{T_p}(t)\}$, the model is unable to capture this behavior properly (see Fig. 17).

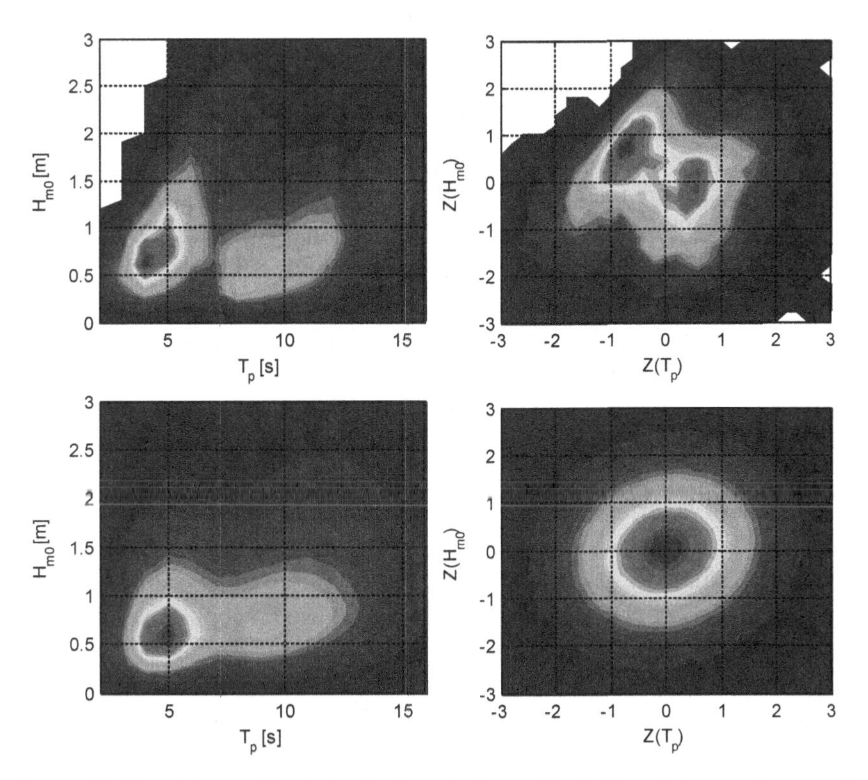

Figure 17. Bivariate empirical distribution of $\mathbf{H_{m0} - T_P}$ (left) and $\mathbf{Z_{H_{m0}} - Z_{T_P}}$ (right) obtained with the original (top) and simulated (bottom) data.

For more details on the evaluation of the VAR models the reader is referred to Solari and van Gelder (2012) and reference therein. In Solari and van Gelder (2012) VAR, TVAR and MSVAR models mentioned on §3.3.1, along with a set of stationary and non-stationary mixture distributions presented on §2.1 and §2.3, are used to simulate 5-variate time series of significant wave height, peak wave period, peak wave direction, wind speed and wind direction.

4.3.2. *Discussion*

It was observed that when using the VAR model for modeling the time dependence structure of the series, most of its behavior can be explained. New simulated series reproduce satisfactorily the auto- and cross-correlation structure of the original series (not shown here; see Solari and van Gelder, 2012), as well as its univariate marginal distributions, and to some extend its interannual variability and its persistence regimes. On the other hand, a remarkable limitation of the VAR model is its inability to successfully reproduce the bivariate behavior of the normalized variables (see Fig. 17 bottom-right), limiting the ability of the model to reproduce the bivariate distribution of the original variables.

Regarding the regime switching VAR models, Solari and van Gelder (2012) found that they are more able in reproducing bivariate behavior of the normalized variables, as well as cases with non-uniform dependence between the variables. This however does not necessarily translate into a significant improvement of agreement between the measured and the simulated bivariate distributions of the original variables.

Two other different approaches that have proven effective for the simulation of bivariate time series are based on the use of copulas. One is the use of a combination of a copula-based Markov model, as described on §3.2.2, and an ARX model (i.e. an Autoregressive model with exogenous variables). Mendonça et al. (2012) used a copula-based model for the simulation of wind speeds, as described on §4.2 and then used an ARX model for the simulation of wind directions, where the exogenous variables was the normalized wind speed. The other approach is based on the use of vine-copulas (see e.g. Kurowicka and Joe, 2011). Solari and Losada (2014) use vine copulas construction approach for the simulation of wind speed and direction time series.

4.4. *Non-Stationary Circular Variables and Inter-Annual Variability: Wave Directions*

In this section a mixture of non-stationary circular distributions is used to analyze the seasonal and inter-annual variability of a series of wave directions. To this end 21 years (1989-2009) of hindcast mean wave directions (θ_m) from the ERA-interim program of the European Centre for Medium-Range Weather Forecasts at coordinates 36°S 52°W (South-American Atlantic coast) are used.

A first exploratory analysis shows that the probability distribution of θ_m at the study site is bimodal, with one peak in direction ENE, other in direction S, and the main mass of the data between both. Moreover, this bimodality is not uniform throughout the year, being the ENE direction predominant during austral summer season and the S direction predominant during austral winter season (see Fig. 18). Additionally, given the location of the case study, it is considered that the climatic indices Niño 3.4 (NINO34), Tropical South Atlantic Index (TSA), Southern Oscillation Index (SOI) and Antarctic Oscillation (AAO or SAM) may have some influence on the inter-annual variability of θ_m. As an example, Fig. 18 shows the empirical PDF of θ_m for all Januaries and for all Julys, along with the PDF considering only month with positive (negative) AAO anomaly.

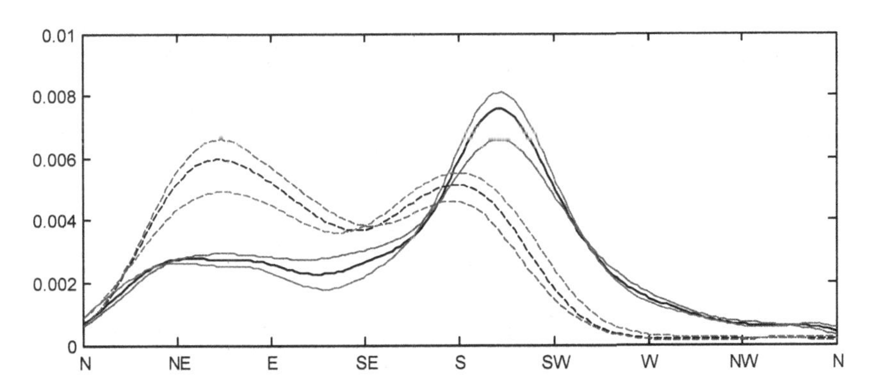

Figure 18. Empirical smoothed PDF of $\boldsymbol{\theta_m}$. Continues lines correspond to Julys and dashed lines to Januaries. Black lines corresponds to the PDF estimated with all Januaries (Julys), red lines correspond to month with AAO<0 and green lines to month with AAO>0.

Given the results obtained with the exploratory analysis, it was decided to implement a mixture of non-stationary Wrapped Normal distributions (see (7) and (8) on §2.1.3) for analyzing the variability of θ_m.

Seasonal variability within each of the parameters of the distribution was modeled by means of a Fourier series approximation of order two ($N_K = 2$ in (9)), while the possible influence of the indices considered (namely AAO, TSA, SOI and NINO34) was included by means of a linear combination of the monthly anomaly of each index; e.g. the term $f_i(C_i(t), t)$ in (10) reduces to a linear relation $\beta_i C_i(t)$, being $C_i(t)$ the monthly anomaly of the indices and β_i a parameter to be fitted. In order to reduce the total number of parameters of the model, only those β values that are significantly different from zero, according to the likelihood ratio test (see e.g. Coles 2001), were retained. Monthly anomaly of the indices were provided by the National Oceanic and Atmospheric Administration of the USA (NOAA Earth System Research Laboratory).

Figure 19 shows the superposition of the empirical and modeled annual non-stationary distribution function. The agreement between the two distributions is noticeable, with the proposed model correctly reproducing the time evolution of the two modes of the empirical distribution.

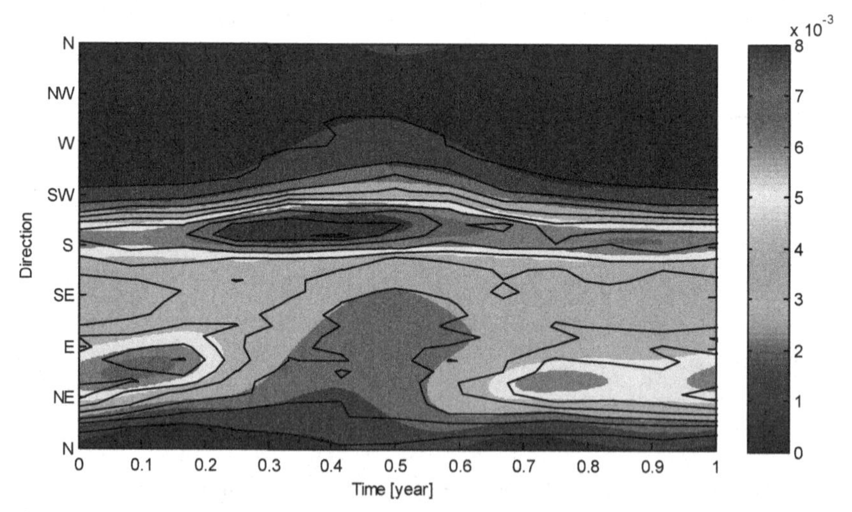

Figure 19. Superposition of the empirical (black lines) and modeled (color filled contours) annual non-stationary distribution function (PDF).

In order to evaluate the importance that each climatic index has on the interannual variability the R coefficient was defined: $R = \beta_i \sigma_i / \Delta_{seasonal}$; i.e. the R coefficient is the ratio between the variation imposed to a given parameter by an anomaly equal to the standard deviation of the whole series of anomalies, and the range of the mean seasonal cycle of the parameter. Table 5 shows the R coefficient estimated from the retained β values (those that are significant different from zero). It is noticed that the indices with the greater influence are Antarctic Oscillation (AAO) and the Tropical South Atlantic Index (TSA). Also, most of the influence is exerted over the WN distribution that represents the southern mode of the distribution (over the waves coming from the south).

Table 5. Mean value for each parameter and ratio between and the amplitude of the annual cycle for each climatic index (Rβ). Only parameters that passed hypothesis testing.

Parameter	Mean value	AAO	TSA	SOI	NINO34
μ1	193	- 0.17	---	---	---
σ12	17	---	1.46	---	---
α1	0.21	- 0.18	---	0.08	---
μ2	56	0.19	---	- 0.03	---
σ22	25	---	---	---	---
α2	0.17	---	---	---	---
μ3	139	- 0.09	0.24	---	---
σ32	68	---	0.31	---	---

Lastly, the model is used to estimate how it would be the distribution of the mean wave directions given a set of values of the climatic indices. In particular, the distribution is estimated assuming strong positive and negative anomalies of the Antarctic Oscillation Index. Fig. 20 shows the mean annual non-stationary distribution obtained assuming zero AAO anomaly (left) and assuming a positive anomaly equal to 1.28 times the standard deviation of the historically observed anomalies (right).

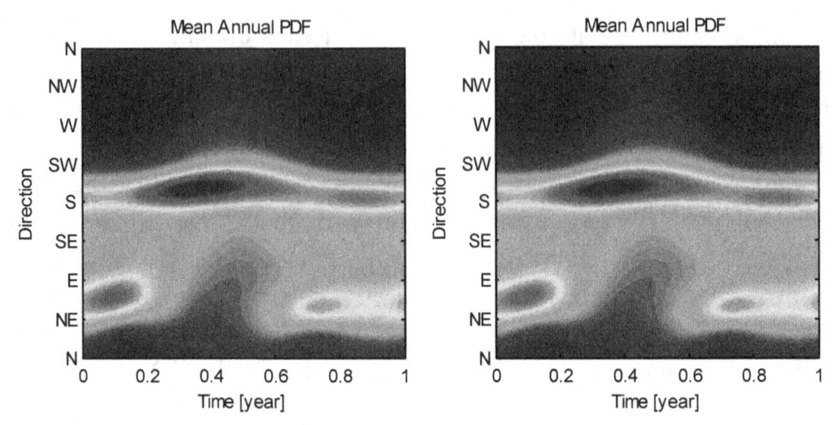

Figure 20. Annual non-stationary distribution function (PDF) considering AAO anomaly equal to zero (left) and equal to 1.28 times the standard deviation of the anomalies (right).

In summary, it is shown that mixtures of non-stationary circular probability distributions are capable of reproducing the distribution of the mean wave directions, including special features of the distribution like bimodality and dependence on climatic indices. Also, by retaining only statistically significant parameters, the final number of fitted parameters may be significantly reduced and the interpretation of the results simplified. For more details about this example and a preliminary analysis of the consequence of directional variability on longshore sediment transport rates, the reader is referred to Solari and Losada (2012c).

5. Epilogue

The world socioeconomic progress is exhausting the energy, the water and coastal zone resources. To overcome this trend a more serious and rigourous slogan must drive the progress today: (1) socioeconomic and environmental progress must be concomitant, (2) use of basic resources must be minimized, and (3) safety, serviceability and operationality of the human works must be maximized (Losada et al., 2009). However harbor and coastal engineering must deal with the environmental events and their random nature. Thus the response to the problem has to include the associated uncertainty, the probability of failure and operational stoppage and the quantification of the consequences (risk assessment).

Losada et al. (2009) formulated and summarized new design principles and tools, based on risk analysis and decision theory, to be applied to coastal and harbor design and management problems. It was only the start, and it was said: "it is a long way before the maritime works are properly optimized and their uncertainty bounded". Among others, the validity and field of application of the overall methodology was questioned in terms of,

a) Data processing: uncertainty related with application of POT method
b) Synthetic database of the wave climate obtained from simulation process.

The work of Solari and Losada (2011, 2012a, 2012b, 2014) can assist in a more representative and precise statistical description of the variables and their temporal simulation including seasonal and interannual variability, for both uni- as well as multi-dimensional random variables. For certain, the new tools are not too friendly to apply, but it is the penalty for taking into account the increasing complexity of the wave climate temporal variability. This variability has effects on the socioeconomic and environmental processes, and on the adaptation to the sea level rise associated with the global warming. Again, it is necessary to be prudent, present work is only a step more; there are still many aspects that have to be formulated, solved and embodied in maritime structure´s design tree.

5.1. *Downscaling the Agents Simulation on the Safety Margin*

The objective of the project design is to verify that the subset of the structure fulfills the project requirements in each and every one of the states. Thus, it is necessary to establish a verification equation for each failure or stoppage mode. These are state equations formed by a set of terms, each one formed by a combination of mathematical operations of parameters, agents, and actions. Generally speaking, the verification equation is applied with the hypothesis that, during the sea state, the outcomes of the terms are stationary and uniform from a statistical point of view.

Terms can be classified as favorable or unfavorable depending on their contribution to the prevention or the occurrence of the failure or stoppage mode, respectively. The safety margin S is defined by the difference between favorable and unfavorable terms. They are expressed according to the state variables, and thus, the probability of failure or stoppage of the structure against the mode depends on the temporal evolution of the values of the terms. As a result, it is necessary to calculate the joint probability function of the safety margin of each failure and stoppage modes in the useful life.

The calculation sequence includes: (1) the determination of the joint density and distribution functions of the agents and the simulation of time series during the useful life of the structure; (2) the determination of the joint density and distribution functions of the terms of the equation; (3) simulation of the safety margins of the modes and calculation of the joint probability of failure and the operativity of the structure. Most of the tools described in this chapter can be applied to simulate the time series of the terms (actions) of the equation, and of the safety margin of each failure and stoppage mode during the useful life.

To bring down the uncertainty of the wave climate (agents) on the safety margins of failure and stoppage modes is an immense task that requires appropriated methods and tools. Moreover, the verification equation itself carry on uncertainty and in many cases, perhaps in too many, it is not quantified. A first attempt to develop such a method is described in Benedicto et al. (2004), and applied to the Harbor of Gijon (Spain). It includes both the downscaling of the wave climate (agents) to the wave forces (actions) and to the safety margin. To reduce the computational effort some simplified methods are proposed (Camus et al., 2011) by selecting a subset of representative sea states and applying a maximum dissimilarity algorithm and the radial basis function. A mixture model similar to that proposed by Solari and Losada, 2011, could help efficiently to choose the representative sea states without increasing the computational effort.

5.2. *Risk Assessment and Decision Makers*

While environmental conditions make a stretch of coast attractive, the activity and decisions of the economic agents interact with the physical environment, altering the littoral processes and the coastal evolution. This interaction may have significant environmental and economic consequences that need to be evaluated a priori. Risk assessment is the proper tool to face these situations. Simulation techniques, as described above, can be used to assess the uncertainty in the performance of a set of predefined investment and management strategies of the coast or the harbor based on different criteria representing the main concerns of interest groups. This way to handle the problem should help the decision makers to propose alternatives and to analyze their performance on different time scales and in terms of economic, social and environmental criteria.

There are different approaches to solving conflicts in which several criteria have to be accounted for, as the Multiple Criteria Decision-Aiding, MCDA method (Mendoza and Martins, 2006) or the Stochastic Multicriteria Acceptability Analysis (SMAA) method (Lahdelma and Salminen, 2001). This second method is especially suited for dealing with situations in which criteria values and/or the preferences of decision makers are unknown. This methodology can be regarded as a negotiation analysis. It was applied as a management solution for Playa Granada in the Guadalfeo River Delta, Spain, where the construction of a dam is causing severe coastal erosion. Coupling the analysis of the uncertainty with a SMAA provides the probability of alternatives obtaining certain ranks and the preferences of a typical decision maker who supports an alternative (Felix et al., 2012).

The methods and tools presented in this chapter are an important step to achieve to the final goal of the new engineering paradigm: to provide the decision makers with the risk assessment of different alternatives and strategies by evaluating the probability of the unfulfillment of the project objectives and requirements and their consequences. There are still many elements of the procedure, which have to be developed, but as showed in this chapter, current advances in knowledge provide an important support to the engineers. Undoubtedly, they improve the traditional engineering methods applied in maritime engineering.

Acknowledgments

The findings and developments described in this chapter were carried out within the framework of the following research projects: Spanish Ministry of Science and Innovation research project CTM2009-10520; Andalusian Regional Government research project P09-TEP-4630; Spanish Ministry of Public Works projects CIT-460000-2009-21 and 53/08-FOM/3864/2008.

Sebastián Solari acknowledge financial support provided by the Spanish Ministry of Education through its postgraduate fellowship program FPU, grant AP2009-03235, and by the Uruguayan Agency for Research and Innovation (ANII) under the post- doctoral scholarship "Fondo Profesor Dr. Roberto Caldeyro Barcia"(PD-NAC-2012-1-7829). Miguel A. Losada acknowledge financial support provided by the Ministerio de Economía y Competitividad through the research project BIA2012-37554 "Método unificado para el diseño y verificación de los diques de abrigo".

The authors also wish to thank Puertos del Estado, Spain, and the European Centre for Medium-Range Weather Forecasts (ECMWF) for providing the reanalysis wind and wave data series used in this chapter.

References

1. Abegaz, F., Naik-Nimbalkar, U., 2008a. Modeling statistical dependence of Markov chains via copula models. J. Stat. Plan. Inference 138, 1131–1146. doi:10.1016/j.jspi.2007.04.028
2. Abegaz, F., Naik-Nimbalkar, U. V., 2008b. Dynamic Copula-Based Markov Time Series. Commun. Stat. - Theory Methods 37, 2447–2460. doi:10.1080/03610920801931846
3. Athanassoulis, G.A., Stefanakos, C.N., 1995. A nonstationary stochastic model for long-term time series of significant wave height. J. Geophys. Res. 100, 16149–16162.

4. Baquerizo, A., M. A. Losada, 2008. Human interactions with large scale coastla morphological evolution. An assessment of uncertainty. Coastal Engineering 55, 569-580. Doi:10.1016/j.coastaleng.2007.10.004.
5. Behrens, C.N., Freitas Lopes, H., Gamerman, D., 2004. Bayesian analysis of extreme events with threshold estimation. Stat. Modelling 4, 227–244. doi:10.1191/1471082X04st075oa
6. Benedicto, I., Castillo, M.C., Moyano, J.M., Baquerizo, A., Losada, M.A. 2004. Verificacion de un tramo de obra de la ampliacion de un puerto mediante la metodologia reomendada en la ROM 0.0. EROM 00, edited by J.R. Medina. Puertos del Estado, Spain, www.puertos.es. (in spanish).
7. Borgman, L.E., Scheffner, N.W., 1991. Simulation of time sequences of wave height, period, and direction.
8. Cai, Y., Gouldby, B., Hawkes, P., Dunning, P., 2008. Statistical simulation of flood variables: incorporating short-term sequencing. J. Flood Risk Manag. 1, 3–12. doi:10.1111/j.1753-318X.2008.00002.x
9. Callaghan, D.P., Nielsen, P., Short, A., Ranasinghe, R., 2008. Statistical simulation of wave climate and extreme beach erosion. Coast. Eng. 55, 375–390. doi:10.1016/j.coastaleng.2007.12.003
10. Camus, P., Mendez, F.J., Medina, R., 2011. A hybrid efficient method to downscale wave climate to coastal areas. Coastal Engineering, 58(9). Elsevier B.V., pp. 851-862 doi:10.1016/j.coastaleng.2011.05.007.
11. Castillo, E., 1987. Extreme value theory in engineering. Academic Press, Inc. pp 389.
12. Coles, S., 2001. An Introduction to Statistical Modeling of Extreme Values (Springer series in statistics). Springer-Verlang.
13. Cunnane, C., 1979. A Note on the Poisson Assumption in Partial Duration Series Models. Water Resour. Res. 15, 489–494.
14. Davison, A.C., Smith, R., 1990. Models for exceedances over high thresholds. J. R. Stat. Soc. Ser. B. 52, 393–442.
15. De Michele, C., Salvadori, G., Passoni, G., Vezzoli, R., 2007. A multivariate model of sea storms using copulas. Coast. Eng. 54, 734–751. doi:10.1016/j.coastaleng.2007.05.007
16. de Waal, D.J., van Gelder, P.H.A.J.M., Nel, A., 2007. Estimating joint tail probabilities of river discharges through the logistic copula. Environmetrics 18, 621–631. doi:10.1002/env
17. de Zea Bermudez, P., Kotz, S., 2010. Parameter estimation of the generalized Pareto distribution—Part II. J. Stat. Plan. Inference 140, 1374–1388.
18. Duan, Q., Gupta, V., Sorooshian, S., 1993. Shuffled complex evolution approach for effective and efficient global minimization. J. Optim. Theory Appl. 76.
19. Duan, Q., Sorooshian, S., Gupta, V., 1992. Effective and efficient global optimization for conceptual rainfall-runoff models. Water Resour. Res. 28, 1015–1031.
20. Dupuis, D.J., 1998. Exceedances over High Thresholds : A Guide to Threshold Selection. Extremes 261, 251–261.

21. Eastoe, E.F., Tawn, J.A., 2012. Modelling the distribution of the cluster maxima of exceedances of subasymptotic thresholds. Biometrika 99, 43–55. doi:10.1093/biomet/asr078
22. Fawcett, L., Walshaw, D., 2006. Markov chain models for extreme wind speeds. Environmetrics 17, 795–809. doi:10.1002/env.794
23. Fawcett, L., Walshaw, D., 2012. Estimating return levels from serially dependent extremes. Environmetrics 23, 272–283. doi:10.1002/env.2133
24. Felix, A., Baquerizo, A., Santiago, J.M., Losada, M.A., 2012. Coastal zone management with stochastic multi-criteria analysis. Journal of Environmental Management 112, 252-266.
25. Fisher, N.I., 1993. Statistical analysis of circular data. Cambridge University Press.
26. Furrer, E.M., Katz, R.W., 2008. Improving the simulation of extreme precipitation events by stochastic weather generators. Water Resour. Res. 44, 1–13. doi:10.1029/2008WR007316
27. Genest, C., Favre, A., 2007. Everything you always wanted to know about copula modeling but were afraid to ask. J. Hydrol. Eng. 12, 347–368.
28. Guanche, Y., Mínguez, R., Méndez, F.J., 2013. Climate-based Monte Carlo simulation of trivariate sea states. Coastal Engineering 80, 107–121. DOI: 10.1016/j.coastaleng.2013.05.005
29. Guedes Soares, C., Cunha, C., 2000. Bivariate autoregressive models for the time series of significant wave height and mean period. Coast. Eng. 40, 297–311. doi:10.1016/S0378-3839(00)00015-6
30. Guedes Soares, C., Ferreira, A.M., 1996. Representation of non-stationary time series of significant wave height with autoregressive models. Probabilistic Eng. Mech. 11, 139–148.
31. Guedes Soares, C., Ferreira, A.M., Cunha, C., 1996. Linear models of the time series of significant wave height on the Southwest Coast of Portugal. Coast. Eng. 29, 149–167.
32. Izaguirre, C., Méndez, F.J., Menéndez, M., Luceño, A., Losada, I.J., 2010. Extreme wave climate variability in southern Europe using satellite data. J. Geophys. Res. 115, 1–13. doi:10.1029/2009JC005802
33. Joe, H., 1997. Multivariate Models and Dependance Concepts (Monographs on Statistics and Applied Probability 73). Chapman & Hall/CRC.
34. Kottegoda, N., Rosso, R., 2008. Applied Statistics for Civil and Environmental Engineers, Second. ed. Blackwell Publishing.
35. Kurowicka, D., Joe, H. (Eds.), 2011. Dependence modeling. Vine copula handbook. World Scientific Publishing Co.
36. Lahdelma, R., Salminen, P., 2001. Smaa-2: stochastic multicriteria acceptability analysis for group decision making. Operating Research 49 (3) 444-454.
37. Leadbetter, M.., 1991. On a basis for "Peaks over Threshold"modeling. Stat. Probab. Lett. 12, 357–362.

38. Losada, M.A., Baquerizo A., Ortega-Sanchez, M., Santiago, J.M., Sanchez-Badorrey, E., 2009. Socioeconomic and environmental risk in coastal and ocean engineering. Chapter 33, Section 8, in Handbook of Coastal and Ocean Engineering. Edited by Y. C. Kim.
39. Luceño, A., Menéndez, M., Méndez, F.J., 2006. The effect of temporal dependence on the estimation of the frequency of extreme ocean climate events. Proc. R. Soc. A 462, 1683–1697.
40. Lütkepohl, H., 2005. New Introduction to Multiple Time Series Analysis. Springer-Verlag.
41. Madsen, H., Rasmussen, P.F., Rosbjerg, D., 1997. Comparison of annual maximum series and partial duration series methods for modeling extreme hydrologic events 1. At-site modeling. Water Resour. Res. 33, 747–757.
42. Mazas, F., Hamm, L., 2011. A multi-distribution approach to POT methods for determining extreme wave heights. Coast. Eng. 58, 385–394. doi:10.1016/j.coastaleng.2010.12.003
43. Méndez, F.J., Menéndez, M., Luceño, A., Losada, I.J., 2006. Estimation of the long-term variability of extreme significant wave height using a time-dependent Peak Over Threshold (POT) model. J. Geophys. 111, 1–13. doi:10.1029/2005JC003344
44. Méndez, F.J., Menéndez, M., Luceño, A., Medina, R., Graham, N.E., 2008. Seasonality and duration in extreme value distributions of significant wave height. Ocean Eng. 35, 131–138.
45. Mendonça, A., Losada, M.A., Solari, S., Neves, M.G., Reis, M.T., 2012. INCORPORATING A RISK ASSESSMENT PROCEDURE INTO SUBMARINE OUTFALL PROJECTS AND APPLICATION TO PORTUGUESE CASE STUDIES, in: Proceedings of 33rd Conference on Coastal Engineering, Santander, Spain. doi:10.9753/icce.v33.management.18
46. Mendoza, G.A., Martins, H., 2006. Multi-criteria decision analysis in natural resource mangement: a critical review of methods and new modelling paradigms. Forest Ecology and Management 230, 1-22
47. Monbet, V., Ailliot, P., Prevosto, M., 2007. Survey of stochastic models for wind and sea state time series. Probabilistic Eng. Mech. 22, 113–126. doi:10.1016/j.probengmech.2006.08.003
48. Nai, J.Y., van Beek, E., van Gelder, P.H.A.J.M., Kerssens, P.J.M.A., Wang, Z.B., 2004. COPULA APPROACH FOR FLOOD PROBABILITY ANALYSIS OF THE HUANGPU RIVER DURING BARRIER CLOSURE, in: Proceedings of the 29th International Conference on Coastal Engineering. pp. 1591–1603.
49. Nelsen, R.B., 2006. An Introduction to Copulas (Springer series in statistics), Second. ed. Springer.
50. Ochi, M.K., 1998. Ocean waves. The stochastic approach. Cambridge Ocean Technology Series 6. Cambridge University Press.
51. Payo, A., Baquerizo, A., Losada, M.A., 2008. Uncertainty assessment: Application to the shoreline. J. Hydraul. Res. 46, 96–104.

52. Pickands, J., 1975. Statistical Inference Using Extreme Order Statistics. Ann. Stat. 3, 119–131.
53. Ribatet, M., Ouarda, T.B.M.., Sauquet, E., Gresillon, J., 2009. Modeling all exceedances above a threshold using an extremal dependence structure: Inferences on several flood characteristics. Water Resour. Res. 45, 1–15. doi:10.1029/2007WR006322
54. ROM 0.0, 2001. General procedure and requirements in the design of harbor and maritime structures. Project Development: Miguel A. Losada, Universidad de Granada. Puertos del Estado, Spain, www.puertos.es.
55. ROM 1.0, 2009. Recommendations for the project design and construction of breakwaters. Project Development: Miguel A. Losada, Universidad de Graanda. Puertos del Estado, Spain, www.puertos.es.
56. Rosbjerg, D., Madsen, H., Rasmussen, P.F., 1992. Prediction in partial duration series with generalized pareto-distributed exceedances. Water Resour. Res. 28, 3001–3010.
57. Ruggiero, P., Komar, P.D., Allan, J.C., 2010. Increasing wave heights and extreme value projections: The wave climate of the U.S. Pacific Northwest. Coast. Eng. 57, 539–552. doi:10.1016/j.coastaleng.2009.12.005
58. Salvadori, G., De Michele, C., Kottegoda, N., Rosso, R., 2007. Extremes in Nature. An Approach Using Copulas (Water Science and Technology Library 56). Springer.
59. Scheffner, N.W., Borgman, L.E., 1992. Stochastic time-series representation of wave data. J. Waterw. Port, Coastal, Ocean Eng. 118, 337–351.
60. Schwarz, G., 1978. Estimating the Dimension of a Model. Ann. Stat. 6, 461–464.
61. Scotto, M.G., Guedes Soares, C., 2000. Modelling the long-term time series of significant wave height with non-linear threshold models. Coast. Eng. 40, 313–327. doi:10.1016/S0378-3839(00)00016-8
62. Serinaldi, F., Grimaldi, S., 2007. Fully Nested 3-Copula: Procedure and Application on Hydrological Data. J. Hydrol. Eng. 12, 420–430.
63. Smith, J.A., 1987. Estimating the upper tail of flood frequency distributions. Water Resour. Res. 23, 1657–1666.
64. Smith, R.L., Tawn, J.A., Coles, S., 1997. Markov chain models for threshold exceedances. Biometrika 84, 249–268.
65. Solari, S., Losada, M.A., 2011. Non-stationary wave height climate modeling and simulation. J. Geophys. Res. 116, 1–18. doi:10.1029/2011JC007101
66. Solari, S., Losada, M.A., 2012a. Unified distribution models for met-ocean variables: Application to series of significant wave height. Coast. Eng. 68, 67–77. doi:10.1016/j.coastaleng.2012.05.004
67. Solari, S., Losada, M.A., 2012b. A unified statistical model for hydrological variables including the selection of threshold for the peak over threshold method. Water Resour. Res. 48, 1–15. doi:10.1029/2011WR011475
68. Solari, S., Losada, M.A., 2012c. PARAMETRIC AND NON-PARAMETRIC METHODS FOR THE STUDY OF THE VARIABILITY OF WAVE

DIRECTIONS : APPLICATION TO THE ATLANTIC URUGUAYAN COASTS, in: Proceedings of 33rd Conference on Coastal Engineering, Santander, Spain. doi:10.9753/icce.v33.waves.14

69. Solari, S., Losada, M.A., 2014. SIMULATION OF NON-STATIONARY WIND SPEED AND DIRECTION TIME SERIES FOR COASTAL APPLICATIONS, in: 34th International Conference on Coastal Engineering (Accepted).
70. Solari, S., van Gelder, P.H.A.J.M., 2012. On the use of Vector Autoregressive (VAR) and Regime Switching VAR models for the simulation of sea and wind state parameters, in: Guedes Soares et al. (Ed.), Marine Technology and Engineering. Taylor & Francis Group, London, pp. 217–230.
71. Stefanakos, C.N., 1999. Nonstationary stochastic modelling of time series with applications to environmental data. National Technical University of Athens.
72. Stefanakos, C.N., Athanassoulis, G.A., 2001. A unified methodology for the analysis, completion and simulation of nonstationary time series with missing values, with application to wave data. Appl. Ocean Res. 23, 207–220.
73. Stefanakos, C.N., Athanassoulis, G.A., 2003. Bivariate Stochastic Simulation Based on Nonstationary Time Series Modelling, in: Proceedings of the 13th International Offshore and Polar Engineering Conference. pp. 46–50.
74. Stefanakos, C.N., Athanassoulis, G.A., Barstow, S.F., 2006. Time series modeling of significant wave height in multiple scales, combining various sources of data. J. Geophys. Res. 111, 1–12. doi:10.1029/2005JC003020
75. Stefanakos, C.N., Belibassakis, K.A., 2005. Nonstationary stochastic modelling of multivariate long-term wind and wave data, in: Proceedings of the 24th International Conference on Offshore Mechanics and Arctic Engineering. pp. 1–10.
76. Tancredi, A., Anderson, C., O'Hagan, A., 2006. Accounting for threshold uncertainty in extreme value estimation. Extremes 9, 87–106.
77. Thompson, P., Cai, Y., Reeve, D., Stander, J., 2009. Automated threshold selection methods for extreme wave analysis. Coast. Eng. 56, 1013–1021. doi:10.1016/j.coastaleng.2009.06.003
78. Vaz De Melo Mendes, B., Freitas Lopes, H., 2004. Data driven estimates for mixtures. Comput. Stat. Data Anal. 47, 583–598. doi:10.1016/j.csda.2003.12.006
79. Walton, T.L., Borgman, L.E., 1990. Simulation of Nonstationary, Non-Gaussian Water Levels on Great Lakes. J. Waterw. Port, Coastal, Ocean Eng. 116, 664–685.

Appendix A: Maximum Likelihood Estimation of Mixture Distributions

Parameters of the mixtures distributions presented in this chapter were estimated by means of Maximum Likelihood. For minimizing the log-likelihood function a global optimization procedure is used, entitled the shuffled complex evolution (SCE-UA) method (Duan et al., 1993, 1992).

In the cases were the original data was truncated and one or more parameters of the distribution were the thresholds, as is the case in distribution (1) on §2.1.1, the convergence of the optimization algorithm was improving by distributing the original data uniformly in their corresponding symmetrical intervals – e.g. hindcast significant wave height data is truncated with a precision of 0.1 m, then the original data series was added a noise with distribution $\mathcal{U}(-0.05, 0.05)$.

When dealing with the non-stationary distribution (§2.3), the estimation of the parameters was perform by progressively increasing the order of approximation of the Fourier series on (9). The parameters obtained for order $n\ (\theta_{a0}, \theta_{a1}, \theta_{b1}, \dots, \theta_{an}, \theta_{bn})$ are the first approximation used to estimate those in order $n+1$, with zero used as the first approximation of the new parameters $(\theta_{an+1}, \theta_{bn+1}) = (0,0)$.

Appendix B: Copula-Based Second- and Third-Order Markov Models

Variables $F_{1|2}$ and $F_{3|2}$ are calculated using the bivariate copula C_{12} that defines the first-order Markov process:

$$F_{1|2} = Prob[P(t) \leq u | P(t-1) = v] = \frac{\partial C_{12}}{\partial v}(u, v) \quad \text{(B1a)}$$

$$F_{3|2} = Prob[P(t-2) \leq w | P(t-1) = v] = \frac{\partial C_{23}}{\partial v}(v, w) \quad \text{(B1b)}$$

where it is assumed that the time-dependence structure is stationary, and thus $C_{12} \equiv C_{23}$.

If these variables are dependent on each other (a dependence measured with τ_K or ρ_S), a trivariate copula C_{123} is then built that contemplates this dependence and which defines the second-order Markov process

$$C_{123}(u, v, w) = Prob[P(t) \leq u, P(t-1) \leq v, P(t-2) \leq w] \quad \text{(B2)}$$

where marginal distributions C_{12} and C_{23} are given by the copula $C_{12} \equiv C_{23}$, and where marginal C_{13} represents the dependence of $P(t)$ and $P(t-2)$ that is not explained by C_{12}. A copula of this type can be found in [Joe, 1997, chap. 4.5]

$$C_{123}(u,v,w) = \int_{-\infty}^{v} C_{13}\left(F_{1|2}(u,x), F_{3|2}(x,w)\right) F_2(dx) \tag{B3}$$

where C_{13} is fit based on the sample of $F_{1|2}$ and $F_{3|2}$.

Similarly, $F_{1|23}$ and $F_{4|23}$ are calculated using C_{123}

$$\begin{aligned} F_{1|23}(u,v,w) &= Prob[P(t) \le u | P(t-1) = v, P(t-2) = w] \\ &= \frac{\partial^2 C_{123}}{\partial v \partial w} \Big/ \frac{\partial^2 C_{23}}{\partial v \partial w} = \frac{\partial C_{13}}{\partial F_{3|2}} \left(F_{1|2}(u,v), F_{3|2}(v,w)\right) \end{aligned} \tag{B4a}$$

$$\begin{aligned} F_{4|23}(v,w,y) &= Prob[P(t-3) \le y | P(t-1) = v, P(t-2) = w] \\ &= \frac{\partial^2 C_{123}}{\partial u \partial v} \Big/ \frac{\partial^2 C_{12}}{\partial u \partial v} = \frac{\partial C_{24}}{\partial F_{2|3}} \left(F_{2|3}(v,w), F_{4|3}(w,y)\right) \end{aligned} \tag{B4b}$$

where $C_{12} \equiv C_{23} \equiv C_{34}$ and $C_{123} \equiv C_{234}$.

If the dependence between $F_{1|23}$ and $F_{4|23}$, measured with Kendall's τ_K or Spearman's ρ_S, is significant, there is a significant degree of dependence between $P(t)$ and $P(t-3)$ that is not explained by C_{123}, and copula C_{1234} is built, which defines the third-order Markov process

$$\begin{aligned} &C_{1234}(u,v,w,y) \\ &\quad = Prob[P(t) \le u, P(t-1) \le v, P(t-2) \le w, P(t-3) \le y] \\ &\quad = \int_{-\infty}^{w} \int_{-\infty}^{v} C_{14}\left(F_{1|23}(u,x_1,x_2), F_{4|23}(x_1,x_2,y)\right) C_{23}(dx_1, dx_2) \end{aligned} \tag{B5}$$

where copula C_{14} is fit, based on the sample of variables $F_{1|23}$ and $F_{4|23}$.

The distribution of $P(t)$ conditioned to $P(t-1) = v$, $P(t-2) = w$ and $P(t-3) = y$ is then obtained by deriving (B6)

$$C_{1|234}(u,v,w,y)$$
$$= Prob[P(t) \le u | P(t-1) = v, P(t-2) = w, P(t-3) = y]$$
$$= \frac{\partial^3 C_{1234}}{\partial v \partial w \partial y} \Big/ \frac{\partial^3 C_{234}}{\partial v \partial w \partial y} = \frac{\partial C_{14}}{\partial F_{4|23}} \left(F_{1|23}(u,v,w), F_{4|23}(v,w,y) \right) \tag{B6}$$

Appendix C: Simulation Procedure of the Third-Order Markov Process

For the third-order Markov process the simulation procedure is:
(i) At $t = 1$, $u_1 \sim \mathcal{U}(0, 1)$ is simulated, and $P_1 = u_1$ is taken.
(ii) For $t = 2$, $u_2 \sim \mathcal{U}(0,1)$ is simulated, and P_2 is calculated conditioned to P_1, solving (C1)

$$u_2 = C_{2|1}(P_1, P_2) \tag{C1}$$

(iii) For $t = 3$, $u_3 \sim \mathcal{U}(0,1)$ is simulated, and P_3 is calculated conditioned to P_1 and P_2, solving (C2)

$$u_3 = C_{3|1}\left(C_{1|2}(P_1, P_2), C_{3|2}(P_2, P_3) \right) \tag{C2}$$

(iv) for $t \ge 4$, $u_t \sim \mathcal{U}(0,1)$ is simulated, and P_t is calculated conditioned to P_{t-1}, P_{t-2} and P_{t-3}, solving (C3)

$$u_t = C_{4|1}\Big(C_{1|23}\left(C_{1|2}(P_{t-3}, P_{t-2}), C_{3|2}(P_{t-2}, P_{t-1}) \right), C_{4|23}\left(C_{2|3}(P_{t-2}, P_{t-1}), C_{4|3}(P_{t-1}, P_t) \right) \Big) \tag{C3}$$

(v) Once the series $\{P_t\}$ is simulated, the series of the original variable $\{V_t\}$ is constructed, using the inverse of the non-stationary mixture distribution function.

In steps (ii) to (iv), the expressions of the conditioned copulas are analytically resolved, whereas equations (C1), (C2) and (C3) are numerically solved.

Printed in the United States
By Bookmasters